Finite State Machine Datapath Design, Optimization, and Implementation

Finite State Machine Datapath Design, Optimization, and Implementation
Justin Davis and Robert Reese

ISBN: 978-3-031-79775-0   paperback
ISBN: 978-3-031-79776-7   ebook

DOI: 10.1007/978-3-031-79776-7

A Publication in the Springer series

SYNTHESIS LECTURES ON DIGITAL CIRCUITS AND SYSTEMS #14

Lecture #14
Series Editor: Mitchell Thornton, Southern Methodist University

Series ISSN

ISSN 1932-3166   print
ISSN 1932-3174   electronic

Finite State Machine Datapath Design, Optimization, and Implementation

Justin Davis
Raytheon Missile Systems

Robert Reese
Mississippi State University

SYNTHESIS LECTURES ON DIGITAL CIRCUITS AND SYSTEMS #14

ABSTRACT

Finite State Machine Datapath Design, Optimization, and Implementation explores the design space of combined FSM/Datapath implementations. The lecture starts by examining performance issues in digital systems such as clock skew and its effect on setup and hold time constraints, and the use of pipelining for increasing system clock frequency. This is followed by definitions for latency and throughput, with associated resource tradeoffs explored in detail through the use of dataflow graphs and scheduling tables applied to examples taken from digital signal processing applications. Also, design issues relating to functionality, interfacing, and performance for different types of memories commonly found in ASICs and FPGAs such as FIFOs, single-ports, and dual-ports are examined. Selected design examples are presented in implementation-neutral Verilog code and block diagrams, with associated design files available as downloads for both Altera Quartus and Xilinx Virtex FPGA platforms. A working knowledge of Verilog, logic synthesis, and basic digital design techniques is required. This lecture is suitable as a companion to the synthesis lecture titled Introduction to Logic Synthesis using Verilog HDL.

KEYWORDS:

Verilog, datapath, scheduling, latency, throughput, timing, pipelining, memories, FPGA, flowgraph

Table of Contents

Table of Figures

CHAPTER 1

Calculating Maximum Clock Frequency

The purpose of this chapter is to find the maximum clock frequency and adjusted setup and hold times based on propagation delays for circuits with combinational and sequential gates. This chapter assumes the reader is familiar with digital gates and memory elements such as latches and flip-flops.

1.1 LEARNING OBJECTIVES

After reading this chapter, you will be able to perform the following tasks:

- Discover the longest combinational delay path through a circuit
- Calculate the three types of delays in sequential circuits
- Calculate chip-level setup and hold time based on internal registers
- Calculate board-level clock frequencies

1.2 GATE PROPAGATION DELAY

The simplest metric of performance of a digital device is computation time. Often this is measured in computations per second and depends on the type of computation. For general-purpose processors, it may be measured in millions of instructions per second (MIPS). For arithmetic processors, it may be measured in millions of floating point operations per second (MFLOPS). Computation time is based partly on the speed of the clock and partly on the number of clocks per operation. This chapter will focus on computing the maximum clock speed to enable the minimum computation time.

A digital logic gate is constructed from transistors arranged in a specific way to perform a mathematical operation. These transistors are operated like on/off switches. Ideally the transistors can switch on to off or off to on instantly; however, realistic transistors have a finite switching time. A leading factor in transistor switching time is their physical size. Smaller transistors will usually switch faster than large transistors. As transistor size is further miniaturized through emerging technologies, this delay continues to decrease. Modern transistors can switch exceptionally fast, but the delay must still be accounted for.

Specific types of transistors in a logic gate are not as important as their effect. The switching delay of the transistors creates a delay in the logic gate. The latter can be measured from the time an input changes to the time an output changes. This delay is called the ***propagation delay***(t_{pd}). This book will only consider the delays associated with the gate but with the understanding that it is defined by the underlying transistors.

1.2.1   Single Input/Multiple Input Delays

The simplest gate for discussing t_{pd} is the inverter. The inverter has one input and one output. While the input is a logic high, the output is a logic low. When the input changes from high to low, the output will change from low to high after a certain delay. The input and the output of the inverter do not change instantaneously from a logic low to a logic high or vice versa. These finite rise times and fall times are shown in Fig. 1.1. The 50% point on the rise time or fall time is when the voltage level is halfway between the logic high and logic low. The t_{pd} is measured between the 50% point of the input rise time and the 50% point of the fall time of the output.

The t_{pd} can be different for the output rise time and fall time. If the rise time is longer than the fall time, then the 50% point will be shifted, which results in a larger t_{pd}. Since the propagation delay can be different, each is denoted differently. When the output is changing from high to low, the delay associated with it is denoted t_{phl}. When the output is changing from low to high, the delay associated with it is denoted t_{plh}. For simplicity, the worst case is taken for the two propagation delays and is considered to be the total t_{pd} for the entire gate.

Even though each type of logic gate is constructed differently, the delay through the gates are measured the same. A multiple input gate has many more propagation delays. For example, an AND gate has at least two inputs as shown in Fig. 1.2. The t_{pd} must be measured from low to high and high to low for each input.

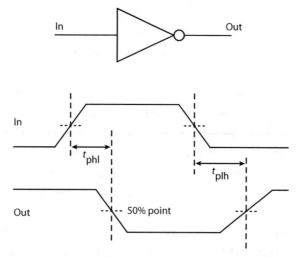

FIGURE 1.1: Inverter propagation delay.

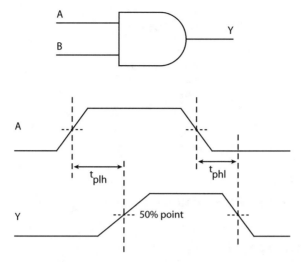

FIGURE 1.2: AND gate propagation delay.

For a two-input gate, four propagation delays are found: $A2Y\_t_{plh}$, $A2Y\_t_{phl}$, $B2Y\_t_{plh}$, $A2Y\_t_{phl}$. For simplicity, the worst case is taken for the four propagation delays and is considered to be the total t_{pd} for the entire gate (Y_t_{pd}). This is true for any number of inputs for a combinational gate. Typically, datasheets for a logic device contains the worst-case t_{pd} along with the typical t_{pd}.

1.2.2 Propagation Delay Effects

When multiple gates are connected together, the propagation delays on the individual gates can produce unwanted and incorrect results in the output called ***glitches***. The glitches can cause output values that are logically impossible with ideal logic gates. For example, an AND gate only outputs a logic high when both inputs are logic high. When the inputs to an AND gate are always opposite as in Fig. 1.3, then the output will never be logic high. If the inverter has a finite t_{pd}, then the output of the AND gate can become a logic high while the signal is propagating through the inverter. When the input X is a logic low, the output of the inverter is a logic high. When the input switches to a logic high, both the inputs to the AND gate are logic high because the change has not propagated through the inverter yet.

Because of propagation delays, whenever multiple gates are combined, the output could have glitches until after all the signals have propagated through all the gates. The output cannot be considered valid until after this delay. This is the reason why digital systems are usually clocked. The rising edge of the clock signifies when all the input signals are sent to the circuit. If the clock period is set correctly, by the time the next rising edge occurs, the glitches end and the output is considered valid. The clock period is set by analyzing all the propagation delays in the circuit.

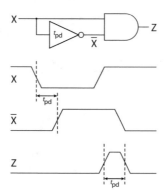

FIGURE 1.3: Glitches caused by propagation delay.

1.2.3   Calculating Longest Delay Path

The t_{pd} for a circuit is found by tracing a path from one input to the output. The propagation delay of each gate is added to the total delay for that path. This procedure is repeated for every path from each input to the output. After a set of all delays is constructed, t_{pd} for the circuit is chosen to be the largest delay in the set.

1.2.4   Example 1.1

An XOR gate can be constructed using AND, OR, and NOT gates as in Fig. 1.4. Using the circuit in Fig. 1.4 and the delays of the AND, OR, and NOT gates in Table 1.1, what is the worst-case t_{pd} for the entire circuit?

For the XOR gate, there are four individual paths from the input to the output. The first path starts at the X input and progresses through the A1 AND gate and the O2 OR gate. The total delay is $25 + 20 = 45$ ns. The second path from the X input progresses through the O1 OR gate, the N3 NOT gate, and the O2 OR gate for $20 + 10 + 20 = 50$ ns delay.

The Y input also has two paths. The first is through the N2 NOT gate, the A1 AND gate, and the O2 OR gate for a $10 + 25 + 20 = 55$ ns delay. The last path is through the N1 NOT gate, the O1 OR gate, the N3 NOT gate, and the O2 OR gate for a $10 + 20 + 10 + 20 = 60$ ns delay. All paths are listed in Table 1.2 .

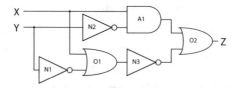

FIGURE 1.4: XOR gate architecture.

TABLE 1.1: Propagation delays for individual gates

Gate	Propagation Delay
NOT	10 ns
AND	25 ns
OR	20 ns

TABLE 1.2: Total set of all propagation delays

Starting Input	Path	Delay
X	A1 + O2	45 ns
X	O1 + N3 + O2	50 ns
Y	N2 + A1 + O2	55 ns
Y	N1 + O1 + N3 + O2	60 ns

The worst-case delay path is 60 ns. On the datasheet, the maximum t_{pd} would be listed as 60 ns. This is also the minimum period of the clock if the XOR gate is used in a real circuit.

1.2.5   Propagation Delays for Modern Integrated Circuits

Delay values for an integrated circuit are dependent upon the technology used to fabricate the integrated circuit, and the environment that the integrated circuit functions within (voltage supply level, temperature). The delays used in this chapter and the next are not meant to reflect actual delays found in modern integrated circuits since those delays are moving targets. Instead, the delay values used in these examples are chosen primarily for ease of hand calculation. The ns unit (nanoseconds,1.0e−9 s) was chosen because nanoseconds is convenient for describing off-chip delays as well as on-chip delays. Furthermore, using a real time unit such as ns instead of unit-less delays allows frequency calculations with real units. See Section 1.6 for a short discussion of how propagation delays for integration circuits have varied as integrated circuit fabrication technology has improved.

1.3   FLIP-FLOP PROPAGATION DELAY

Flip-flops and latches are considered memory elements because they can output a set value without an input. This value can be changed as needed. The input is transferred to the output when the device is enabled. In this book, a flip-flop will be defined by the enable (usually a clock) being an

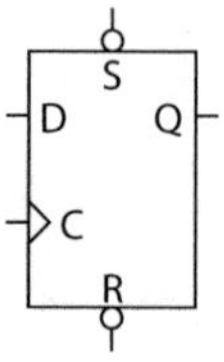

FIGURE 1.5: D-type flip-flop input options.

edge-triggered signal. For a latch, the enable is a level-sensitive signal. This book uses flip-flops in its examples since this is the most commonly-used design style. While many types of flip-flops exist such as SR flip-flops, D flip-flops, T flip-flops, or JK flip-flops, this book will only discuss D flip-flops since they are the simplest and most straight-forward. The other types of flip-flops can be analyzed using the same techniques as the D flip-flop. In D flip-flops, the input is copied to the output at the clock edge. The D flip-flop can have a variety of input options as shown in Fig. 1.5.

A specialized type of flip-flop is called a register. Registers have an enable input which prevents the latter from being transferred to the output in every clock cycle. The input will only be copied when the enable is set high. Registers can come in arrays, which all have the same control signals, but have different data inputs/outputs. Sometimes the term register is used synonymously with the term flip-flop.

The output for a memory element has a t_{pd} like a combinational gate; however, it is measured differently. Since the output for a register only changes on a clock transition, t_{pd} is measured from the time the clock changes to the time the input is copied to the output. Since the data output does not change when the data input changes, t_{pd} is not measured from the data input to the data output. However, the clock-to-output propagation delay (t_{C2Q}) is not the only delay associated with a register.

1.3.1   Asynchronous Delay

Other inputs are available for different types of registers. Some registers have the ability to be set to a logic high or reset to a logic zero from independent inputs. These set/reset inputs can take effect either on a clock edge or independent of the clock altogether. When an input is dependant on the clock edge, it is called a *synchronous* input. When an input is not dependant on the clock, it is called an *asynchronous* input. The data input to a register is always a synchronous input. An asynchronous set-to-output delay is labeled (t_{S2Q}) and an asynchronous reset-to-output delay is labeled (t_{R2Q}). If the set/reset inputs are synchronous, then there are no individual delays associated with them since the clock-to-output delay covers their delay. Other inputs are available for registers such as an enable input, but again any input, which is dependant on the clock, will not have a separate propagation delay.

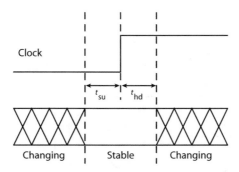

FIGURE 1.6: Relative setup and hold time timing.

1.3.2   Setup and Hold Time

Registers have an additional constraint to ensure that the input is correctly transferred to the output. For every synchronous input, the signal must remain at a stable logic level for a set amount of time before the clock edge occurs. This is called the **setup** (tsu) time for the register. Additionally, the input signal must remain stable for a set amount of time after the clock edge occurs. This is called the **hold** (thd) time for the register. If the input changes within the setup or hold time, then the output cannot be guaranteed to be correct. This specification is indicated on the datasheet for the register and is set by the characteristic of the internal transistors. Fig. 1.6 illustrates setup and hold time concepts.

1.4   SEQUENTIAL SYSTEM DELAY

Most digital systems contain both sequential and combinational circuits. These circuits can be more difficult to analyze for the longest delay path. Three different types of delay paths occur in the circuit. Each delay path is analyzed differently depending on the origin and destination of the path. The first type of path starts at the data or control inputs to the circuit and is traced through to the outputs of the circuit passing through only combinational gates. This is called a pin-to-pin propagation delay. The next type of path starts at the clock input and is traced to the outputs of the circuit passing through at most one register. This is called t_{C2Q}. The last type of path starts at a register and is traced to another register. This is called the register-to-register delay.

1.4.1   Pin-to-Pin Propagation Delay

A pin-to-pin propagation delay path (t_{P2P}) is defined by any path from an input to an output that passes through only combinational gates, which means **it cannot pass through any registers**. This is similar to Section 1.2.3 when the longest delay path was found through multiple combinational gates. A path is formed from the input to the output and all of the gate delays are added together. This is repeated for all possible combinational paths. It is possible there are no paths from the input to the output that contain only combinational gates. In this case, t$_{P2P}$ does not contribute to finding the minimum clock period.

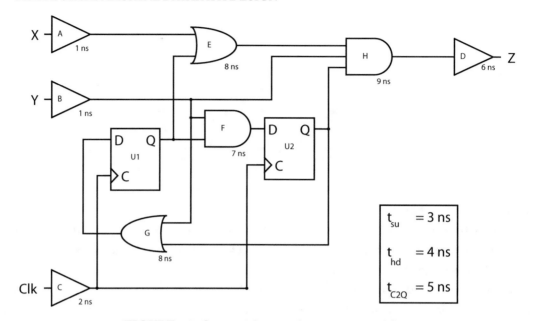

FIGURE 1.7: Sequential circuit for propagation delay.

1.4.2   Example 1.2

The circuit in Fig. 1.7 is the internal layout of a custom built chip. The t_{pd} for each gate is listed below it. The delays for the register are all the same and listed in the lower right corner. Input protection circuits and output fan-out circuitry can slow down the signal transmission on and off the chip. These delays will be represented as simple buffers on the schematic. Find t_{P2P}.

There are multiple pin-to-pin combinational paths for this circuit. The inputs X and Y both have combinational-only paths to the output. The clock (Clk) input does not have a combinational-only path to the output because any path would pass through one of the two registers.

For input X, the path starts at the input buffer A and proceeds through the OR gate E, the AND gate H, and the output buffer D. The propagation delays for these gates are added together to get $1 + 8 + 9 + 6 = 24$ ns.

$$A\_t_{pd} + E\_t_{pd} + H\_t_{pd} + D\_t_{pd} = t_{P2P} \tag{1.1}$$

$$1 + 8 + 9 + 6 = 24 \text{ ns} \tag{1.2}$$

For the input Y, the path starts at the input buffer B and proceeds through the AND gate H, and the output buffer D. The propagation delays for these gates are added together to get $1 + 9 + 6 = 16$ ns.

$$B\_t_{pd} + H\_t_{pd} + D\_t_{pd} = t_{P2P} \tag{1.3}$$

TABLE 1.3: Total set of all pin-to-pin propagation delays

Starting Input	Path	Delay
X	A + E + H + D	24 ns
Y	B + H + D	16 ns

$$1 + 9 + 6 = 16 \, \text{ns} \qquad (1.4)$$

The larger of these two delays is the worst-case t_{P2P} for this circuit. The path "A + E + H + D" is the worst-case with a delay of 24 ns. The list of delays is in Table 1.3.

1.4.3   Clock-to-Output Delay

The second type of t_{pd} path is the clock-to-output path (t_{C2Q}). These paths pass through *exactly one register*. The clock input is routed to the registers in the circuit. A path is traced from the clock input of the system to the clock input of a register. Then the path continues through that register to the output of the circuit. The delays of the combinational gates along the path and the clock-to-output delay of the register are added to the total delay of the path.

Often two clock-to-output delays exist when analyzing a circuit. One is for the internal registers, and the other is for the entire circuit. The register C2Q will be a part of the system C2Q, so the register C2Q will always be the smaller of the two. The combinational delay before the register is listed as t_{comb_I2C}, and the combinational delay after the register is listed as t_{comb_Q2O}.

$$t_{comb\_I2C} + t_{C2Q\_FF} + t_{comb\_Q2O} = t_{C2Q\_SYS} \qquad (1.5)$$

Some circuit analysis programs treat the clock-to-output delay the same as the pin-to-pin combinational delay, so sometimes on the analysis report there will be no clock-to-output delay listed. The clock input is counted as a regular input. Often these reports will list the worst-case delays for each input, so the clock-to-output delay can be found by searching this list.

1.4.4   Example 1.3

Using the same circuit in Fig. 1.7, find the worst-case t_{C2Q}.

There are two clock-to-output paths through the circuit. Both paths pass through the input buffer C. One path then proceeds through the first register U1, through the OR gate E, through the 3-input AND gate H, and finally to the output buffer D.

$$C\_t_{pd} + U1\_t_{C2Q} + E\_t_{pd} + H\_t_{pd} + D\_t_{pd} = t_{C2Q\_SYS} \qquad (1.6)$$

$$2 + 5 + 8 + 9 + 6 = 30 \, \text{ns} \qquad (1.7)$$

TABLE 1.4: Total Set of all clock-to-output propagation delays		
Starting Input	**Path**	**Delay**
Clk	$C + U1 + E + H + D$	30 ns
Clk	$C + U2 + H + D$	22 ns

The second path proceeds through the second register U2, through the 3-input AND gate H, and finally to the output buffer D.

$$C\_t_{pd} + U2\_t_{C2Q} + H\_t_{pd} + D\_t_{pd} = t_{C2Q\_SYS} \tag{1.8}$$

$$2 + 5 + 9 + 6 = 22\,\text{ns} \tag{1.9}$$

The larger of these two delays is the worst-case t_{C2Q} for this circuit. The path "C + U1 + E + H + D" is the worst-case with a delay of 30 ns. The list of delays is in Table 1.4.

1.4.5   Register-to-Register Delay

The last type of propagation delay is the register-to-register delay (t_{R2R}). This is usually the largest of the three types of delays in modern circuit designs. Consequently, it is usually the delay that sets the minimum clock period. As the name of this delay path suggests, this delay path starts at the output of a register and is traced to the input of another register. The path could even be traced back to the input of the starting register, but the route always involves *at most two registers.* The number of register-to-register paths in a circuit is proportional to the number of registers in the design. Specifically, the number of paths will be at most 2^N where N is the number of registers. Therefore, the number of paths that must be checked can increase very quickly as a design grows.

The t_{R2R} must be equal to or larger than the clock period. At the beginning of the clock period, the clock transitions from low to a high. This change propagates through the register for a fixed amount of time before the input is transferred to the output. This is the clock-to-output delay of the register. Once the input is present on the output, the combinational gates after the output will begin to switch. After the changes propagate through the combinational gates, the new signals will be ready at the inputs to the registers for transfer to the outputs of the registers. Furthermore, the new signals must satisfy the setup time of the register to ensure they will be transferred correctly to the output.

$$t_{C2Q\_FF} + t_{comb\_R2R} + t_{su\_FF} = t_{R2R} \tag{1.10}$$

TABLE 1.5: Total set of all register-to-register propagation delays

Starting Input	Path	Delay
U1	U1 + F + U2	15 ns
U2	U2 + G + U1	16 ns

1.4.6   Example 1.3

Using the same circuit in Fig. 1.7, find the worst-case t_{R2R}

There are two registers in this design. Starting with register U1, there is only one path from the output of this register to another register. This path passes through gate F to the input of register U2. Therefore, computing this register-to-register path is easy.

$$U1\_t_{C2Q} + F\_t_{pd} + U2\_t_{su} = t_{R2R} \tag{1.11}$$

$$5 + 7 + 3 = 15\,\text{ns} \tag{1.12}$$

Starting with register U2, there is only one path from the output to another register. This path passes through gate G to the input of register U1.

$$U2\_t_{C2Q} + G\_t_{pd} + U1\_t_{su} = t_{R2R} \tag{1.13}$$

$$5 + 8 + 3 = 16\,\text{ns} \tag{1.14}$$

The two register-to-register paths in Table 1.5 above are 15 ns and 16 ns. The worst-case t_{R2R} is therefore 16 ns through the path "U2 + G + U1". If all the registers have the same clock-to-output delay and t_{su} (as is often the case), the only difference between the paths is the combinational circuits between the registers. This can make computing t_{R2R} much easier.

1.4.7   Overall worst-case delay

Now that the maximum delays for the three types of paths have been found, the overall maximum delay of the sequential system can be found. The worst case is the largest delay of the three path types. For the example circuit in Fig. 1.7, the three worst cases are listed in Table 1.6.

The worst-case delay for this system is the clock-to-output delay at 30 ns. Therefore, for this sequential system, the minimum clock period is 30 ns in order to allow all gate outputs to reach stable values. This corresponds to a maximum clock frequency of 33.3 MHz.

1.4.8   Setup and hold adjustments

An additional requirement for sequential circuits is to ensure that t_{su} and t_{hd} requirements of the internal registers have been met. Signals external to the circuit must not violate t_{su} before the clock

TABLE 1.6: Total set of worst-case propagation delays

Path Type	Path	Delay
P2P	A + E + H + D	24 ns
C2Q	C + U1 + E + H + D	30 ns
R2R	U2 + G + U1	16 ns

and t_{hd} after the clock *at the inputs to the internal register.* If the sequential circuit was going to be packaged into a chip and sold to a customer, the customer may not know how to check if the internal register setup and hold requirements have been met. Therefore t_{su} and t_{hd} requirements are recomputed for the entire sequential circuit and that information is passed to the customer.

For setup time, the data signal must not change for a given time before the clock edge. If the input signal is delayed, such as, through a combinational gate or input buffer as in Fig. 1.8, the input may violate the t_{su} requirement. Therefore, any delay added between the input pin and the register input must be added to the setup time requirement. The delay between the clock input pin and the clock input to the register must also be subtracted from t_{su} . This means if the delays between the pins to the register are the same, there will be no change in t_{su}. Only when there is a difference in the delays will the setup time change.

This procedure must be repeated for each register in the design that has an external input routed to its input through any combinational path. The longest delay from the data input to the registers is used as the worst case. The shortest delay from the clock input to the registers is used as the worst case. The difference between these two paths is the adjustment to the setup time.

$$(t_{pd\_data(MAX)} - t_{pd\_clk(MIN)}) + t_{su\_FF} = t_{su\_TOTAL} \tag{1.15}$$

For hold time, if the clock signal is delayed, such as through an input buffer, the input may violate the t_{hd} requirement. The worst case for t_{hd} is the opposite worst case for t_{su} : the longest delay from the clock input of the circuit to the register, and the shortest delay from the data input to the register. The difference between these two paths is the adjustment to the hold time.

$$(t_{pd\_clk(MAX)} - t_{pd\_data(MIN)}) + t_{hd\_FF} = t_{hd\_TOTAL} \tag{1.16}$$

FIGURE 1.8: Calculating adjusted setup/hold times.

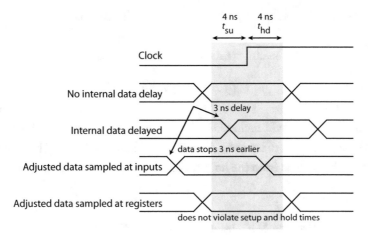

FIGURE 1.9: Adjusted setup and hold timings.

When t_{su} and t_{hd} have been adjusted correctly for the external inputs, the internal t_{su} and t_{su} at the register inputs will not be violated. The timing diagram in Fig. 1.9 shows the behavior of internal delays, which can cause changes in the setup and hold requirement.

1.4.9   Example 1.4   Using the same circuit in Fig. 1.7, find the adjustments to the t_{su} and t_{hd} for the circuit.

In this design, the data input is delivered to the input to two registers. The first path is routed from the Y input through the input buffer, through the OR gate G, and then to the input of the U1 register. The second path passes through the input buffer, through the AND gate F, and then to the U2 register. Note there are no paths from the X input to the inputs of any registers. Table 1.7 provides the set of all input to register delays.

The calculation for t_{su} will include the longest data delay and the shortest clock delay. For this example, the longest data delay is $t_{pd\_data\_U1}$ that will add 9 ns to t_{su}. The shortest clock delay is

TABLE 1.7: Total set of all input to register delays

Delay Path	Path	Delay	Path Name
Y to U1	B + G + U1	9 ns	$t_{pd\_data\_U1}$
Y to U2	B + F + U2	8 ns	$t_{pd\_data\_U2}$
Clk to U1	C + U1	2 ns	$t_{pd\_clk\_U1}$
Clk to U2	C + U2	2 ns	$t_{pd\_clk\_U2}$

$t_{\text{pd\_clk\_U1}}$ that will subtract 2 ns from t_{su}. Given t_{su} of 3 ns, the external t_{su} for this circuit is 10 ns.

$$(t_{\text{pd\_data\_U1}} - t_{\text{pd\_clk\_U1}}) + t_{\text{su\_FF}} = t_{\text{su\_TOTAL}} \tag{1.17}$$

$$(9 - 2) + 3 = 10\,\text{ns} \tag{1.18}$$

The calculation for t_{hd} will include the longest clock delay and the shortest data delay. For this example, the longest clock delay is $t_{\text{pd\_clk\_U1}}$ that will add 2 ns to t_{hd}. The shortest data delay is $t_{\text{pd\_data\_U1}}$ that will subtract 8 ns from the hold time. Given t_{hd} of 4 ns, the external t_{hd} for this circuit is -2 ns.

$$(t_{\text{pd\_clk(MAX)}} - t_{\text{pd\_data(MIN)}}) + t_{\text{hd\_FF}} = t_{\text{hd\_TOTAL}} \tag{1.19}$$

$$(2 - 8) + 4 = -2\,\text{ns} \tag{1.20}$$

The setup and hold window is 8 ns in which the data cannot change. The negative sign in the hold time calculation means the data input can actually start changing before the clock signal. This is not an intuitive behavior for a digital circuit, so often a negative t_{hd} will be specified as zero instead. By setting t_{hd} to zero, the effective setup and hold window has increased to 10 ns.

1.5   BOARD-LEVEL TIMING CALCULATION

A digital chip will usually be used in a larger system connected to other chips. Even if all chips in the system may be rated to operate at a specific clock frequency, the entire system may not.

1.5.1   Datasheet compilation

The datasheet of each chip should have all of the relevant timing information to compute the board-level maximum clock frequency. This data is similar to the gate delays when computing the chip-level maximum clock frequency. Six relevant pieces of data are needed to ensure the operation of the board-level system. The maximum clock frequency of each chip must be provided since the board-level system cannot operate faster than that. The t_{su} and t_{hd} must be provided to ensure no write violation to the registers internal to the chip. The combinational delay and clock-to-output delay must be known to compute the maximum clock frequency of the circuit. The needed information is presented in Table 1.8 along with the values for the example results.

Each chip can be treated as a sequential circuit with both synchronous and asynchronous delays much like a register. Each of the three worst-case delay path types can be computed with the above information to find the maximum clock frequency. The maximum clock frequency for the board will never exceed any individual chip's rating listed on the datasheet.

Parameter	Description	Min	Max	Units
TABLE 1.8: Datasheet for the chapter example				
Tclk	Clock Period	26		ns
Fclk	Clock Frequency		33.3	MHz
t_{su_Y}	Y Setup Time	10		ns
t_{hd_Y}	Y Hold Time	0		ns
$X\_t_{pd\_P2P}$	Combinational delay	24		ns
t_{pd_C2Q}	Clock-to-output delay	30		ns

1.5.2 Board-level maximum frequency

The procedure to find the maximum clock frequency at the board-level is same as at the chip level. The worst-case delays must be found in three cases: the pin-to-pin combinational, the clock-to-output and the register-to-register delays. The minimum clock period is set to the largest of these three paths or the minimum clock period for each individual chip.

1.5.3 Example 1.5

Using the circuit in Fig. 1.10, find the maximum clock frequency. Each chip is the circuit in Fig. 1.7 and uses the timings in Table 1.6.

First, the pin-to-pin combinational delay is found for any path from the X input to the output. There is one pin-to-pin path from the input A to the X input of U1, to the X input of U2, to the

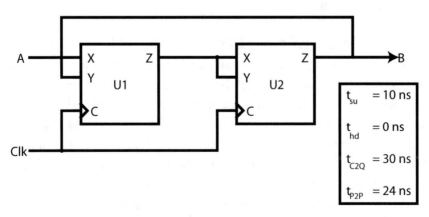

FIGURE 1.10: Board-level schematic to compute maximum clock frequency.

output B. The delay of this path adds the two pin-to-pin delays together $24 + 24 = 48$ ns.

$$X(U1)\_t_{pd} + X(U2)\_t_{pd} = t_{P2P} \tag{1.21}$$

$$24 + 24 = 48 \text{ ns} \tag{1.22}$$

Two clock-to-output delays exist for this circuit. The first path passes through the clock input of U1, through the X input of U2. The second path passes only through the clock input of U2. Since the clock-to-output delays for each chip are the same, the first path will be longer since $30 + 24 = 54$ ns.

$$U1\_t_{C2Q} + X(U2)\_t_{pd} = t_{C2Q\_SYS} \tag{1.23}$$

$$30 + 24 = 54 \text{ ns} \tag{1.24}$$

Three t_{R2R} exist for this circuit. The first path goes through the U1 clock-to-output, through the X input of U2, and then back to the Y input of U1. The second is through the U1 clock-to-output to the input of Y on U2. The third is through the U2 clock-to-output to the input of Y on U1. The longest path is the first since it passes through the combinational portion of U2 for $30 + 24 + 10 = 64$ ns.

$$U1\_t_{C2Q} + X(U2)\_t_{pd} + U1\_t_{su} = t_{C2Q\_SYS} \tag{1.25}$$

$$30 + 24 + 10 = 64 \text{ ns} \tag{1.26}$$

The three worst-case paths and the chip minimum clock period limit the clock frequency for the board-level system. The largest of these values (48 ns, 54 ns, 64 ns, 30 ns) is 64 ns, which is the minimum clock period for the board which corresponds to 15.63 MHz. This frequency is much lower than the chip clock frequency. Note that the combinational delay of the chip contributes most of the slow-down to the circuit.

1.6   DELAYS AND TECHNOLOGY

As stated earlier, delay values for an integrated circuit are dependent upon the technology used to fabricate it, and the environment within which the integrated circuit functions (voltage supply level, temperature). Gate delays for complementary metal-oxide-semiconductor (CMOS) integrated circuits have become smaller over time because transistor channel lengths have become smaller, resulting in transistors that switch faster, and thus, smaller propagation delays for gates. Shrinking transistor sizes have allowed more transistors to be placed in the same integrated circuit, allowing for increased integrated circuit functionality. In programmable logic terms, this means that new generations of programmable logic are able to implement increasing numbers of logic gates in a single package.

TABLE 1.9: Xilinx Virtex FGPA delays over time (delays in picoseconds)

DELAY TYPE	VIRTEX 1 2.5 (2200 NM, V, 1998)	VIRTEX-2 1.5 (1500 NM, V, 2000)	VIRTEX-4 (90 NM, 1.2 V, 2004)	VIRTEX-5 1.0V, (65 NM, 2006)
LUT Propagation Delay	700	390	170	90
DFF Tcq	1200	500	310	40
DFF setup	700	330	400	40
DFF hold	0	−80	−90	20
IOB out (LVTTL)	3200	1510	2020	1520
IOB in (LVTTL)	90	76	87	70

Notes: DFF T_{su}/T_{hd} for Virtex-5 are native setup/hold.
DFF T_{su}/T_{hd} for Virtex 1,2,4 include mux delay.
IOB out/in for Virtex 4,5 uses fast 24mA LVTTL.

Table 1.9 shows delay evolution for the Xilinx Virtex family of field programmable gate arrays (FPGAs) over time. The top row gives each FPGA family name as well as the CMOS technology, supply voltage, and date of first introduction. A CMOS technology designated as 2200 nm (nanometer = 1.0e–9 m) means that the shortest channel MOS transistors has a channel length of 2200 nm (the value 2200 nm is more commonly written as 0.22 μm, but nm is used for consistency purposes). The Xilinx Virtex FPGA family uses a static RAM lookup table (LUT) as the programmable logic element. A LUT is a small memory that is used to implement a boolean function; its contents are loaded from a non-volatile memory at power up. The Virtex 1, 2, and 4 families use a 16×1 LUT, which means that it can implement one boolean function of four variables; the Virtex-5 family uses a 64×2 LUT (two boolean functions of the same six variables). The LUT delays given in Table 1.9 are for a mid-range *speed grade* of these devices. CMOS integrated circuits being made on the same fabrication line can have a range of delays because of variations in the CMOS fabrication process. Thus, devices coming off a fabrication line are tested and separated into different speed grades, with the higher performing devices being sold at a premium price. The supply voltages of Table 1.9 have decreased over time because transistor-switching speeds reach a maximum at lower voltages as transistor channel lengths shrink. Lowering the supply voltage has the added benefit of reducing power consumption, which is important because excessive heating due to high power consumption has become a problem as increasing number of transistors are used in a single integrated circuit.

The delays of Table 1.9 are given in picoseconds (1 ps = 1.0e–12 s). Observe that the LUT propagation delays in Table 1.9 have decreased by almost an order of magnitude across the families (the Virtex-5 LUT t_{pd} would be even faster if it used the smaller LUT of the previous families). The D-flip-flops (DFF) Clock-to-Q propagation delay shows a similar improvement. The DFF t_{su} and t_{hd} are hard to compare because these times include a MUX delay on the D-input of the DFF for the Virtex 1, 2, and 4 families – the setup/hold times for the Virtex-5 DFF does not include this delay. However, in general, DFF t_{su} and t_{hd} also decrease as transistor channel lengths decrease. The Input/Output buffer (IOB) delays are relatively constant over this time because the bonding pad size used to connect the integrated circuit to the package does not shrink as transistor channel length shrinks. The delays associated with any digital logic within the IO pad decreases, but the IO pad delay is dominated by the off-chip load for an output pad, and by the input pad capacitive load for the input pad. Any changes in these delays over time are due to architectural changes in the pad design, such as providing different ranges of output drive strength current, or the need to accommodate different IO standards over time.

For modern programmable logic devices, the device delays are kept in a database that is included in the design toolkit being used to create the design. The timing analysis tool in the FPGA vendor's design toolkit uses these device delay times to calculate external setup and hold times, maximum operating frequency, and internal setup and hold constraints using the timing equations presented in this chapter.

1.7   SUMMARY

This chapter has discussed how to find the important timings of a circuit such as maximum clock frequency by analyzing the delay paths through the gates and registers. By categorizing the delay paths through the circuit, the total number of delay paths that need to be calculated can be minimized. These timings of the internal chip design can also be used to find the maximum clock frequency of the board-level system.

1.8   SAMPLE EXERCISES

For each of the following circuits:

    a. Calculate the worst-case pin-to-pin combinational delay, clock-to-output delay, and register-to-register delay.

    b. Use this data to find the maximum clock frequency.

    c. Calculate t_{su} and t_{hd} for the external inputs.

1.

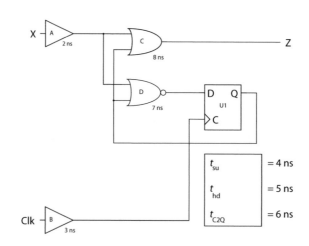

2.

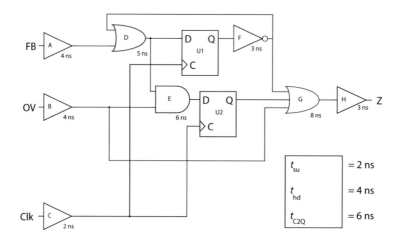

3. Caution, gate E adds a complicating factor!

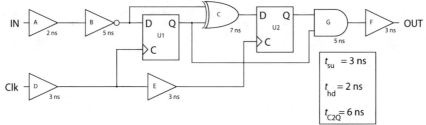

1.9   SAMPLE EXERCISE ANSWERS

1.

Parameter	Calculation	Min	Max	Units
$X\_t_{pd\_P2P}$	$2 + 8 = 10$	10		ns
t_{pd_C2Q}	$3 + 6 + 8 = 17$	17		ns
t_{pd_R2R}	$6 + 7 + 4 = 17$	17		ns
Tclk	max(10, 17, 17)	17		ns
Fclk	1/Tclk		58.8	MHz
t_{su_X}	$4 + (2 + 7) - 3 = 10$	10		ns
t_{hd_X}	$5 + 3 - (2 + 7) = -1$ or 0	0		ns

2.

Parameter	Calculation	Min	Max	Units
$OV\_t_{pd\_P2P}$	$4 + 8 + 3 = 15$	15		ns
t_{pd_C2Q}	$2 + 6 + 3 + 8 + 3 = 22$	22		ns
t_{pd_R2R}	$6 + 3 + 5 + 6 + 2 = 22$	22		ns
Tclk	max(15, 22, 22)	22		ns
Fclk	1/Tclk		45.5	MHz
t_{su_FB}	$2 + (4 + 5 + 6) - 2 = 15$	15		ns
t_{hd_FB}	$4 + 2 - (4 + 5) = -3$ or 0	0		ns

3.

Parameter	Calculation	Min	Max	Units
$IN\_t_{pd\_P2P}$	0	0		ns
t_{pd_C2Q}	$3 + 3 + 6 + 5 + 3 = 20$	20		ns
t_{pd_R2R}	$6 + 7 + 3 + 3$ (gate E) $= 19$	16		ns
Tclk	max(0, 20, 19)	20		ns
Fclk	1/Tclk		50	MHz
t_{su_IN}	$3 + (2 + 5 + 7) - (3 + 3) = 11$	11		ns
t_{hd_IN}	$2 + 3 - (2 + 5) = -2$ or 0	0		ns

CHAPTER 2

Improving Design Performance

The purpose of this chapter is to increase the maximum clock frequency and improve the setup and hold timing by modifying the circuit design. This chapter assumes the reader is familiar with digital gates and memory elements such as latches and registers and can analyze a circuit to find the maximum clock frequency.

2.1 LEARNING OBJECTIVES

After reading this chapter, you will be able to perform the following tasks:

- Maximize the clock frequency by adding output registers
- Minimize the setup and hold window by adding input registers
- Adjust delay measurements when including a delay locked loop (DLL)
- Recalculate the timing of the board-level system after timing modification

2.2 INCREASING MAXIMUM CLOCK FREQUENCY

The three types of delays paths through a circuit set the maximum clock frequency for the design. The only way to increase the maximum clock frequency is to reduce the delay through these worst-case paths. Assuming the propagation delays of the gates and registers cannot be changed, only changing the circuit architecture can reduce the worst-case path delays.

Reducing the worst-case delays by adding circuit elements is not intuitive, but it is effective in increasing performance. For example, the pin-to-pin combinational delay through a circuit can be completely removed by ensuring there are no combinational paths from any input to any output. Likewise, t_{C2Q} can be minimized by reducing combinational paths between the clock input and the output. Both of these tasks can be accomplished by using the same method. Placing registers on all outputs of the circuit removes all combinational delay paths, and minimizes the combinational path of t_{C2Q}.

Adding registers to the design may seem like it would reduce the clock frequency, but in fact it can often increase it. Analyzing the worst-case paths is the only way to set maximum clock frequency. If the worst-case path delay is reduced, then the circuit naturally can be clocked faster. While the pin-to-pin combinational delay is inherently removed from the analysis, the clock-to-output is usually reduced to its minimum possible value. Since the registers are placed at the output of the circuit, there are no combinational circuits after this to add to the clock-to-output delay. The only clock-to-output delay paths possible are through these output registers, so the analysis is greatly simplified.

The output registers can only be added before the combinational output buffer delay because this is not an actual gate in the design. This delay represents the interface from the chip to the board. Often the output circuitry design has a significant delay because of the need for a high fan-out, larger voltage swing, and over-voltage protection. Therefore, placing the register immediately before this buffer is the optimum location.

One consequence of this approach is the impact of t_{R2R} through the circuit. Since there are more registers in the design, there are more register-to-register delays to be computed. Sometimes the worst-case t_{R2R} will increase because of this. If the clock frequency is being limited by the pin-to-pin delay or the clock-to-output delay, and then those delays are reduced, the clock frequency will still increase if t_{R2R} is not increased by a significant amount. If registers are added to the outputs, the worst-case t_{R2R} will usually become the largest delay path of the circuit.

Another consequence of this approach is the impact on latency. Latency is the time required for an input to propagate through a circuit to the output. If a circuit is all combinational, then the latency is in the same clock period in which the data input is applied. By adding registers to the output of the circuit, the latency increases into the next clock period. Adding a set of registers to all outputs of a device means the latency of each input will increase to the beginning of the next clock period. While this is a disadvantage, the impact on performance is usually not significant. The latency has increased, but the clock period has decreased as well (usually). Therefore, the combination of these two effects often cancels each other out.

While latency may have increased by one clock cycle, the rate at which data is being input and output is the same. New data is input and output every clock cycle. The throughput of the data is the same, even though the latency has increased. Therefore, the overall computing performance of the device will increase. This effect is called pipelining, which will be covered in much more detail in the next chapter.

2.2.1 Example 2.1

Add a register to the output of the circuit in Fig1.7 and recompute the maximum clock frequency. Compare the new computations with the computations before the circuit improvements. The new circuit is shown in Fig. 2.1.

The analysis for this circuit is the same as for all maximum clock frequency calculations. The worst-case pin-to-pin combinational delay, clock-to-output delay, and t_{R2R} must be found. Since the output is now registered, there is no pin-to-pin combinational delay. This measurement can be excluded from the analysis, or set to zero for continuity in the final comparison.

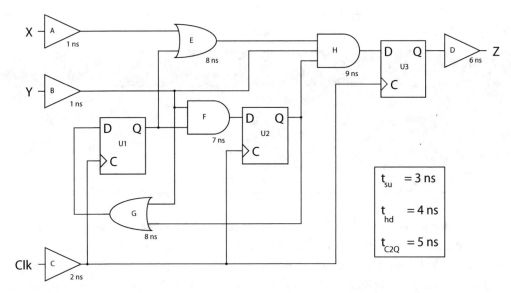

FIGURE 2.1: Adding an output register to the sequential circuit.

The clock-to-output delay only has one path to compute. Since this delay can pass through at most one register, the only register it can now pass through to the output is the new added register. This path proceeds from the clock buffer C, through the register U3, and through the output buffer D. The improved clock-to-output delay is 13 ns.

$$C\_t_{pd} + U3\_t_{C2Q} + D\_t_{pd} = t_{C2Q\_SYS} \tag{2.1}$$

$$2 + 5 + 6 = 13ns \tag{2.2}$$

The number of register-to-register paths has increased due to adding another register from two to four. The paths are listed in Table 2.1 . The worst-case path is from U1, through gates E and H, to the new output register U3 for a total delay of 25 ns.

TABLE 2.1: Total set of new register-to-register propagation delays

Starting input	Path	Delay
U1	U1 + F + U2	15 ns
U2	U2 + G + U1	16 ns
U1	U1 + E + H + U3	25 ns
U2	U2 + H + U3	17 ns

TABLE 2.2: Measured improvement of adding output registers

Measurement	Original delay	Improved delay
P2P	24 ns	0 ns
C2Q	30 ns	13 ns
R2R	16 ns	25 ns
Clock Period	30 ns	25 ns
Clock Frequency	33.3 MHz	40 MHz

The clock period is set by taking the largest of the three worst-case paths, zero ns for the pin-to-pin combinational delay, 13 ns for the clock-to-output delay, and 25 ns for t_{R2R}. Therefore, the minimum clock period is 25 ns, which corresponds to a maximum clock frequency of 40 MHz.

Before adding the register on the output, the minimum clock period was set by the clock-to-output delay. Since this delay decreased to 13 ns, it is no longer limiting the clock period. The t_{R2R} has increased, but is still less than the previous limiting value of 30 ns. This means the maximum clock frequency has significantly increased by adding a single register to the design. The total comparison of measured values is present in Table 2.2.

2.3   IMPROVING SETUP AND HOLD TIMES

Adding registers to the output of the circuit also changes t_{su} and t_{hd} for the circuit. If the circuit has a combinational path through the circuit and a register is added to the output, the longest combinational delay path from a circuit input to a register input could very likely be the newly added register. The setup and hold window could increase significantly because of the new output register. One way to minimize the effects of adding output registers is to place registers on the inputs of the circuit. This will reduce the combinational paths to the registers to minimize the setup and hold window. The input registers can only be placed after the input buffer delay since this is not an actual buffer much like the output buffer delay. Therefore, there will be an input buffer combinational delay to the register input.

2.3.1   Example 2.2

Recompute t_{su} and t_{hd} before and after adding registers to the inputs of the circuit as in Fig. 2.2. This circuit includes the output registers added in the previous example.

The t_{su} of the circuit before adding input registers is computed by finding the longest combinational path to any register in the design. The addition of the output register increases the worst-case delay to 18 ns from the circuit input X to the U3 register through gates A, E, and H. The minimum

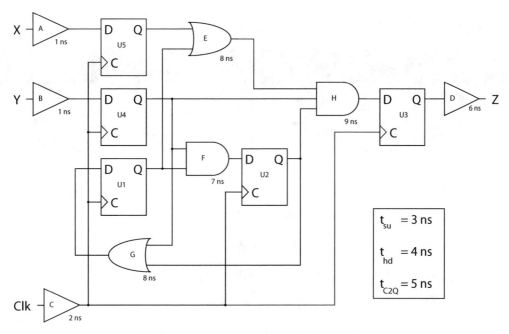

FIGURE 2.2: Adding input registers to the sequential circuit.

clock delay remains the same. Therefore, the new circuit t_{su} increases to 19 ns.

$$(t_{pd\_data\_U1} - t_{pd\_clk\_U1)}) + t_{su\_FF} = t_{su\_TOTAL} \tag{2.3}$$

$$(18 - 2) + 3 = 19\,\text{ns} \tag{2.4}$$

The t_{hd} of the circuit before adding input registers is computed by finding the shortest combinational path to any register in the design. The addition of the output register does not increase this value. The shortest path is the same as the previous analysis at 8 ns. This means t_{hd} remains the same at −2 ns, which should be set to zero since it is negative. The setup and hold window is now 19 ns because of the addition of the output registers.

$$(t_{pd\_clk(MAX)} - t_{pd\_data(MIN)}) + t_{hd\_FF} = t_{hd\_TOTAL} \tag{2.5}$$

$$(2 - 8) + 4 = -2\text{ns} \tag{2.6}$$

Adding input registers after the input buffers simplifies the computations because the number of paths from each input is reduced to one per input. For this circuit, the combinational delay for each input is 1 ns, and the delay for the clock is 2 ns. This means the new t_{su} is 2 ns, and the new t_{hd} is 5 ns. This means the setup and hold window is now 7 ns. The comparison between t_{su} and t_{hd}

TABLE 2.3: Measured improvement of adding input registers

Measurement	Original	Added output registers	Added input registers
Setup Time	10 ns	19 ns	2 ns
Hold Time	0 ns	0 ns	5 ns
Setup and Hold Window	10 ns	19 ns	7 ns

is given in Table 2.3.

$$(t_{\mathrm{pd\_data\_U1}} - t_{\mathrm{pd\_clk\_U1}}) + t_{\mathrm{su\_FF}} = t_{\mathrm{su\_TOTAL}} \qquad (2.7)$$

$$(1 - 2) + 3 = 2\mathrm{ns} \qquad (2.8)$$

$$(t_{\mathrm{pd\_clk(MAX)}} - t_{\mathrm{pd\_data(MIN)}}) + t_{\mathrm{hd\_FF}} = t_{\mathrm{hd\_TOTAL}} \qquad (2.9)$$

$$(2 - 1) + 4 = 5\mathrm{ns} \qquad (2.10)$$

The setup and hold window is nearly doubled when output registers were added to the design. When registers were added to the inputs, the setup and hold window decreased to the smallest possible window. The window cannot decrease below this because it is limited by the setup and hold window of the register, which is also 7 ns.

2.4   DELAY LOCKED LOOPS

Often modern designs that have internal clocks have some type of Phased Locked Loop (PLL) or Delay Locked Loop (DLL) to stabilize and adjust the clock. A PLL is a circuit that creates a completely new clock internal to the circuit, but based on the external clock provided to it. A DLL passes the external clock to the circuit, but adjusts its timing through a network of delays. There are significant differences between these two types of clock management schemes, but they are beyond the focus of this book. For this chapter, the term DLL will be used to describe both PLLs and DLLs. The relevant feature to this material is how DLLs can adjust the phase of the internal clock.

A clock signal can be easily manipulated because of its predictability. The clock will always have a repeating 1-0-1-0 pattern. Therefore, once the clock is active, the clock is the same from one clock period to the next. If the external clock signal is delayed by an input buffer, the internal clock will not be aligned with the external clock. A DLL can artificially make the clock appear to be aligned by inserting additional delay to the clock. For example, an external clock with a period of 8 ns passes through an input buffer that delays the signal by 1 ns as in Fig. 2.3. The DLL measures that the two clocks are not aligned, and then it inserts additional delay to the internal clock until they are aligned. In this example, the DLL would add a 7 ns delay to make the two clocks aligned.

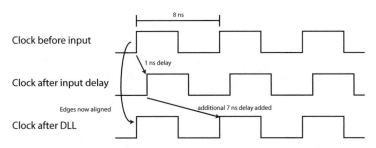

FIGURE 2.3: Operation of a delay locked loop.

A DLL can change the phase of the internal clock either manually or automatically. The advantage of this is that the active clock edge can be placed anywhere. This means the clock delay in the clock-to-output calculations and t_{su} and t_{hd} calculations can be set to whatever needed. Typically the DLL will align the internal clock with the external clock to remove any delays added by the input buffer for the clock signal. The input buffer will add a fixed delay to the clock signal, and the DLL will effectively reduce the delay by that same amount. Note that this technique is not possible to reduce the delays on the data signals because they don't have a predictable repeating pattern.

2.4.1   Example 2.3
Use a DLL to align the internal clock to the external clock in Fig. 2.2. Find any changes to the previous calculations.

Any equation that uses the delay of the input buffer C must be recalculated with that value set to zero. The first change is in the calculation of the clock-to-output delay for the circuit. There is only one clock-to-output path through the circuit through the output register. The new clock-to-output delay for this circuit is reduced by 2 ns to 11 ns.

$$C\_t_{pd} + U3\_t_{C2Q} + D\_t_{pd} = t_{C2Q\_SYS} \tag{2.11}$$

$$0 + 5 + 6 = 11\text{ns} \tag{2.12}$$

The pin-to-pin combinational delay and the register-to-register delay are not affected by the change to the clock because they do not include the clock buffer C. The maximum clock frequency must be checked because this change might affect it if the clock-to-output delay was the limiting factor. Typically t_{R2R} limits the maximum clock frequency, so often the clock frequency will not change when adding a DLL.

The t_{su} and t_{hd} also depend on the clock delay, so they will be affected by adding a DLL. The minimum and maximum clock delay is set to zero and t_{su} and t_{hd} are recalculated.

$$(t_{pd\_data\_U1} - t_{pd\_clk\_U1}) + t_{su\_FF} = t_{su\_TOTAL} \tag{2.13}$$

$$(1 - 0) + 3 = 4\text{ns} \tag{2.14}$$

TABLE 2.4: Datasheet for the improved circuit example

Parameter	Description	Old min	Old max	New min	New max	Units
Tclk	Clock Period	30		25		ns
Fclk	Clock Frequency		33.3		40	MHz
t_{su_Y}	Y Setup Time	10		3		ns
t_{hd_Y}	Y Hold Time	0		4		ns
$X\_t_{pd\_P2P}$	Combinational delay	24		N/A		ns
t_{pd_C2Q}	Clock-to-output delay	30		11		ns

$$(t_{pd\_clk(MAX)} - t_{pd\_data(MIN)}) + t_{hd\_FF} = t_{hd\_TOTAL} \tag{2.15}$$

$$(0 - 1) + 4 = 3\text{ns} \tag{2.16}$$

The new t_{su} is 4 ns, and the new t_{hd} is 3 ns. The setup and hold window has not changed from 7 ns.

2.5   BOARD-LEVEL TIMING IMPACT

The final calculation of the chip is to analyze how well the circuit will improve the board-level performance. The same circuit should be used as in last chapter's example even though the internal design is significantly different. The datasheet for the improved circuit is listed in Table 2.4 . The new calculations include both input and output registers and a DLL for clock adjustment.

2.5.1   Example 2.3

Using the circuit in Figure 2.4, find the maximum clock frequency. Each chip has the same circuit as in Figure 2.2 and uses the timings in Table 2.4.

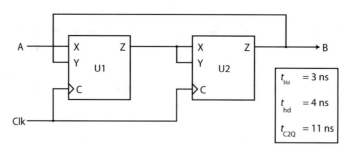

FIGURE 2.4: Board-level schematic to compute maximum clock frequency.

First, since there is no combinational path through the chip, there is no calculation for the pin-to-pin combinational path for the board. This value is excluded when computing maximum clock frequency.

One clock-to-output delay exists for this circuit. This path passes only through the clock input of U2. If there is no clock delay, the clock-to-output for the board is the same as the clock-to-output of the chip. This delay is 11 ns.

Two register-to-register delays exist for this circuit. The first is through the U1 clock-to-output to either input on U2. The third is through the U2 clock-to-output to the input of Y on U1. Both paths have the same delay of $11 + 4 = 15$ ns.

$$U1\_t_{C2Q} + U2\_t_{su} = t_{C2Q\_SYS} \tag{2.17}$$

$$11 + 4 = 15\text{ns} \tag{2.18}$$

The three worst-case paths and the chip minimum clock period limit the clock frequency for the board-level system. The largest of these four values (0 ns, 11 ns, 15 ns, 25 ns) is 25 ns, which is also the minimum clock period for the chip. This means the board can operate at the same frequency as the chips on the board. Note the removal of the combinational paths greatly reduces the delays at the board level.

2.6   SUMMARY

By understanding the parameters that dictate the maximum clock frequency of a circuit, the design can be modified to reduce the longest delays to improve circuit performance. Reducing the combinational delay paths increases the maximum clock frequency by targeting the worst-case paths. By registering all inputs and outputs, the circuit can operate at its maximum frequency within a larger system. Using additional technologies like DLLs can further increase the circuit performance within a larger system.

2.7   SAMPLE EXERCISES

For each of the following circuits, place registers on all data inputs after the input buffer delay and place registers on all data outputs before the output buffer delay. Then,

a) calculate the worst-case pin-to-pin combinational delay, clock-to-output delay, t_{R2R},

b) use this data to find the maximum clock frequency,

c) calculate t_{su} and t_{hd} for the external inputs,

d) determine all effects on the circuit if a DLL was used to remove the clock input buffer delay.

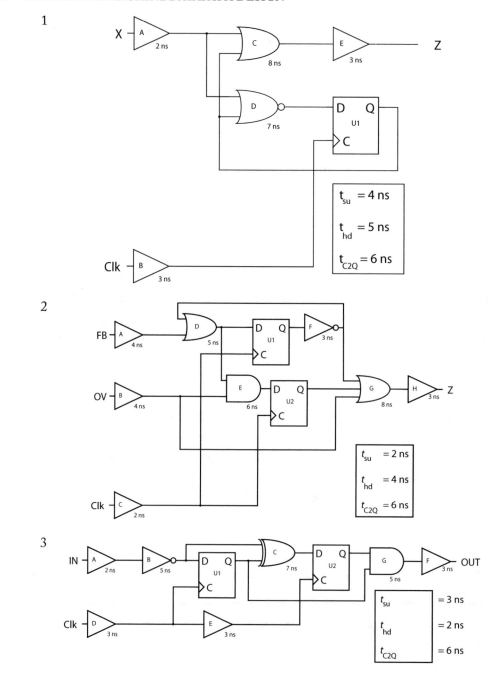

3. For this problem, assume the clock routed to the output register passes through both clock buffers, and the clock to the input register passes through only the first clock buffer, and the DLL only removes the delay in clock buffer D.

2.8   SAMPLE EXERCISE ANSWERS

1.

Parameter	Calculation	Min	Max	Units
$X\_t_{pd\_P2P}$	N/A	0		ns
t_{pd_C2Q}	$3 + 6 + 3 = 12$	12		ns
t_{pd_R2R}	$6 + 8 + 4 = 18$	18		ns
Tclk	max (0, 12, 18)	18		ns
Fclk	1/Tclk		55.6	MHz
t_{su_X}	$4 + 2 - 3 = 3$	3		ns
t_{hd_X}	$5 + 3 - 2 = 6$	6		ns

DLL effects:

Parameter	Calculation	Min	Max	Units
t_{pd_C2Q}	$0 + 6 + 3 = 9$	9		ns
t_{su_X}	$4 + (2-0) = 6$	6		ns
t_{hd_X}	$5 + 0 - 2 = 3$	3		ns

2.

Parameter	Calculation	Min	Max	Units
$OV\_t_{pd\_P2P}$	0	0		ns
t_{pd_C2Q}	$2 + 6 + 3 = 11$	11		ns
t_{pd_R2R}	$6 + 3 + 5 + 6 + 2 = 22$	22		ns
Tclk	max(0, 11, 22)	22		ns
Fclk	1/Tclk		45.5	MHz
t_{su_FB}	$2 + 4 - 2 = 4$	4		ns
t_{hd_FB}	$4 + 2 - 4 = 2$	2		ns

DLL effects:

Parameter	Calculation	Min	Max	Units
t_{pd_C2Q}	$0 + 6 + 3 = 9$	9		ns
t_{su_FB}	$2 + 4 - 0 = 6$	6		ns
t_{hd_FB}	$4 + 0 - 4 = 0$	0		ns

3.

Parameter	Calculation	Min	Max	Units
$IN\_t_{pd\_P2P}$	0	0		ns
t_{pd_C2Q}	$3 + 3 + 6 + 3 = 15$	15		ns
t_{pd_R2R}	$6 + 5 + 7 + 3 - 3 = 18$	18		ns
Tclk	max(0, 15, 18)	18		ns
Fclk	1/Tclk		55.6	MHz
t_{su_IN}	$3 + 2 - 3 = 2$	2		ns
t_{hd_IN}	$2 + 3 - 2 = 3$	3		ns

DLL effects:

Parameter	Calculation	Min	Max	Units
t_{pd_C2Q}	$0 + 3 + 6 + 3 = 12$	12		ns
t_{su_IN}	$3 + 2 - 0 = 5$	5		ns
t_{hd_IN}	$2 + 0 - 2 = 0$	0		ns

CHAPTER 3

Finite State Machine With Datapath Design

This chapter explores finite state machine with datapath (FSMD) design techniques for streaming data applications such as video or audio processing, which can require dedicated logic to meet throughput or latency requirements.

3.1 LEARNING OBJECTIVES

After reading this chapter, you will be able to perform the following tasks:

- Discuss fixed-point representation and saturating arithmetic.
- Transform a streaming data calculation expressed as an equation into a dataflow graph (DFG) format.
- Discuss speed and area tradeoffs in datapath design in relation to latency, throughput, initiation period, and clock period.
- Design datapaths using both non-overlapped/overlapped computations, and non-pipelined/pipelined execution units.
- Design a datapath to implement a DFG that meets target latency and initiation period requirements.

3.2 FSMD INTRODUCTION AND MOTIVATION

A *datapath* contains the components of a digital system that perform the numerical computations for the system. The datapaths described in this chapter perform addition and multiplication on fixed-point numbers with registers used to store intermediate calculations. In this chapter, the generic term of *execution unit* (EU) is used to refer to computation blocks such as adders and multipliers. This chapter uses execution units as black boxes; the reader is referred to a book such as [1] for detailed information on adder and multiplier design.

A finite state machine sequences the computations on the datapath's execution units, with the combined system referred to as FSMD. The FSMD designs in this chapter are tailored to execute a fixed sequence of computations on a dataset. Tradeoffs with regard to the number of required execution units versus the number of clock cycles to complete the computation are studied through

TABLE 3.1: Fixed-format examples

FORMAT	RANGE	EXAMPLES
8.0	0 to 255	143 = 'b10001111; 37 = 'b00100101
5.3	0 to 31.875	17.875 = 'b10001111; 4.625 = 'b00100101
0.8	0 to 0.99609375	0.55859375 = 'b10001111, 0.14453125 = 'b00100101

example implementations. An FSMD approach is used in an application if high performance is needed, as an FSMD implementation typically requires fewer clock cycles than a stored program (a computer) implementation. However, the FSMD logic is fixed, and can only perform its designed computation. A stored program implementation is more flexible, as altering the program that the computer executes modifies the target computation. This is the classic tradeoff of flexibility versus performance when choosing whether to use a stored program or FSMD approach for implementing a digital system. If an application is complex enough, its computations can be divided among cooperating digital systems, with an FSMD handling time-critical computations and a stored program system handling the remaining computations.

A good example of cooperating digital systems is found in a hand-held gaming system, whose task is to execute a game with three-dimensional (3D) graphics. The game application is handled by the microprocessor, while the 3D graphics is performed by a dedicated graphics processor whose core logic is an FSMD optimized for pixel processing. This chapter uses simplified equations from 3D graphics and digital signal processing to illustrate FSMD design tradeoffs.

3.3   FIXED-POINT REPRESENTATION

A *fixed-point* number is a binary number whose format is $X.Y$, where X and Y are the number of binary digits to the left and right of the decimal point, respectively. For unsigned numbers, the integer portion defined by X ranges from 0 to 2^X-1, while the fractional portion ranges from 0 to $1-2^{-Y}$. Table 3.1 gives some examples of eight-bit fixed-point numbers for three different choices of X and Y.

To convert an unsigned decimal number to a $X.Y$ fixed-point format, multiply the decimal number by 2^Y, drop any fractional remainder, and then convert this to its unsigned binary value using a $N.0$ format, where $N = X + Y$. From the 5.3 format example of Table 3.1, the multiplication $4.625 * 2^3 = 37$, which is 0b00100101 as an eight-bit number.

To convert an $X.Y$ unsigned binary number to its decimal representation, first convert the number to its decimal representation assuming an $N.0$ format, where $N = X + Y$. Then divide this number by 2^Y to produce the final decimal result. From the 5.3 format example of Table 3.1, the value 0b10001111 converted to its 8.0 value is 143, which is 17.875 when divided by 2^3.

The numbers in a fixed-point datapath are assumed to share a common *X.Y* format. The logic used to implement binary addition and multiplication works the same regardless of where the decimal point is located, as long as both numbers have the same X.Y format, i.e., the decimal points are *aligned*. This is in contrast to a *floating-point* datapath, which can perform computation on numbers whose decimal points do not align. Floating-point computation blocks require significantly more logic to implement than fixed-point logic blocks. Floating-point computation is used in applications that require an extended range for its numerical data. This chapter does not cover floating-point number encoding or implementation of floating-point computational elements. However, since this chapter treats computation elements as black boxes, the lessons learned in this chapter concerning clock-cycle versus execution unit tradeoffs in datapath design using fixed-point datapaths can easily be applied to floating-point datapaths.

3.4   FIXED-POINT REPRESENTATION IN 3D GRAPHICS

As mentioned previously, 3D graphics is a good example of an application that requires the performance of a dedicated FSMD engine. The *frame rate* of a 3D graphics processor is the number of times per second that a new image is generated for a 3D scene. Each frame is composed of pixels, with a typical resolution being 1280×1024 pixels, or 1,310,270 pixels. The color of each pixel is represented by three eight-bit values that specify the red, green, blue (RGB) color components. Many computations are performed on each RGB component of a pixel to determine the final RGB values of a pixel. Each eight-bit RGB component is a 0.8 fixed-point number. Thus, pixel computations can be thought of as computations on numbers whose range is $[0–1.0)$, which means $0.0 < = c < 1.0$ if c is an RGB component value. From Table 3.1, it is seen that the maximum value of a 0.8 fixed-point number is 0.99609375, which is very close to 1.0. The advantage of the 0.8 fixed-point format is seen in the next section, which discusses saturating arithmetic for fixed point numbers.

3.5   UNSIGNED SATURATING ARITHMETIC AND FIXED-POINT NUMBERS FIXED-POINT REPRESENTATION

Overflow occurs in a computation when the numerical result is outside of the number range supported by a particular data format. A carry out of the most significant bit in an unsigned, fixed-point addition is an overflow indicator. Overflow indicates that the result is incorrect and typically this error condition is handled by the application. However, in real-time data computations such as 3D graphics, video, or audio processing there is no opportunity for the application to correct the error. In these cases, saturating arithmetic is used to saturate the result to the maximum or minimum number in the number range to produce a result that is closer to the correct answer than what overflow produces. Figure 3.1a shows an example of a fixed point addition using normal binary addition that

(a) unsaturating 8-bit addition

	8.0 format	0.8 format
'h50	80	0.3125
+ 'hC0	+ 192	+ 0.75
'h10	16	0.0625

8-bit result has overflowed

(b) saturating 8-bit addition

	8.0 format	0.8 format
'h50	80	0.3125
+ 'hC0	+ 192	+ 0.75
'hFF	255	0.99609375

8-bit result is saturated to maximum value

(c) unsaturating 8-bit subtraction

	8.0 format	0.8 format
'h50	80	0.3125
- 'hC0	- 192	- 0.75
'h90	144	0.5625

8-bit result has underflowed

(d) saturating 8-bit subtraction

	8.0 format	0.8 format
'h50	80	0.3125
- 'hC0	- 192	- 0.75
'h00	0	0.0

8-bit result is saturated to minimum value

FIGURE 3.1: Saturating addition.

overflows, as the result is greater than the maximum value of 255. A saturating adder that clips the result to its maximum value in the overflow case is shown for the same operation in Fig. 3.1b. While the results in Fig. 3.1a and Fig. 3.1b are both incorrect, the saturating operation produces a result that is closer to the correct answer, which is desirable in applications that cannot take any other corrective action on overflow. Figure 3.1c demonstrates an *underflow* case (a borrow into the most significant binary digit) for unsigned eight-bit subtraction. The same operation is performed in Fig. 3.1d using a saturating subtraction operation, which clips the result to its minimum value of zero.

An eight-bit unsigned saturating adder is shown in Fig. 3.2. The output is saturated to its maximum value of 'b11111111 when the eight-bit sum produces a carryout of '1'.

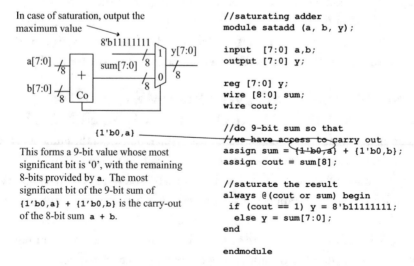

In case of saturation, output the maximum value

This forms a 9-bit value whose most significant bit is '0', with the remaining 8-bits provided by a. The most significant bit of the 9-bit sum of {1'b0,a} + {1'b0,b} is the carry-out of the 8-bit sum a + b.

```
//saturating adder
module satadd (a, b, y);

input   [7:0] a,b;
output  [7:0] y;

reg  [7:0] y;
wire [8:0] sum;
wire cout;

//do 9-bit sum so that
//we have access to carry out
assign sum = {1'b0,a} + {1'b0,b};
assign cout = sum[8];

//saturate the result
always @(cout or sum) begin
  if (cout == 1) y = 8'b11111111;
  else y = sum[7:0];
end

endmodule
```

FIGURE 3.2: Unsigned saturating adder (8-bit).

3.6   MULTIPLICATION

A good question to ask at this point is "How does saturating arithmetic operate for multiplication?" To answer this, recall that the binary multiplication of two N-bit numbers, $N \times N$, requires a $2N$-bit result to contain all of the bits produced by the multiplication. However, it is usually not possible to retain these $2N$-bit in the datapath calculation, as successive multiplications would continually require the datapath size to double in order to prevent any data loss. Assuming that only N bits of an $N \times N$ bit multiplication is kept, then two strategies can be used for discarding half of the bits of the $2N$-bit product. If the fixed-number format used for the calculation is $N.0$ (integers), then a saturating multiplier can be built that saturates the result to the maximum value in case of overflow in the same manner as was done for addition. In this case, the upper N-bit of the $2N$ bit product is discarded and the lower eight-bit saturated to its maximum value.

Another approach is to encode the fixed-point numbers in a $0.N$ format, which means that the product of the $N \times N$ multiplication can never overflow, since the two N-bit numbers being multiplied are always less than one. Hardware saturation of the result is not required; instead, the lower eight-bit of the $2N$-bit product are discarded. The bits that are discarded are the least significant bits of the product, causing successive multiplications to automatically saturate towards a minimum value of zero, as precision is lost due to only retaining eight bits of the product. This will be the approach used in this chapter, as the multiplier design does not have to be modified and the examples used in this chapter assume a 0.8 fixed-point number format.

3.7   THE *BLEND* EQUATION

Equation. (3.1) gives the *blend* equation that is used to illustrate some basic datapath design concepts. The C_{new} value in the blend equation is a new color formed by blending two colors C_a and C_b via a blend factor F. The color values C_{new}, C_a, and C_b are 0.8 fixed-point values whose range is [0–1.0), i.e., $0 \leq C < 1.0$. However, the blend factor F is a nine-bit value encoded to allow the range [0.0–1.0], i.e., $0 \leq F \leq 1.0$. The inclusion of one in the range allows C_{new} to be equal to C_a if F is one, or C_{new} to be equal to C_b if F is zero.

$$C_{new} = C_a \times F + C_b \times (1 - F) \tag{3.1}$$

The nine-bit encoding of F is 'b100000000 if F is equal to one, and $0dddddddd$ for any other value of F, where $dddddddd$ is the 0.8 fixed point equivalent of F. For computation speed purposes, the lower eight-bit of $1-F$ is computed as the one's complement value of the lower eight-bit of F when F is not equal to one or zero. The one's complement operation produces an error of one least significant bit (LSb), but this is deemed acceptable in pixel blend operations, in which computation speed is the most critical factor. The $1-F$ operation implementation is shown in Fig. 3.3. The *mxa* multiplexer and the zero detect logic handle the special case of $F = 0.0$ (0b000000000), in which case the output is 1.0 (0b100000000). The *mxb* multiplexer handles the case of $F = 1.0$, which is

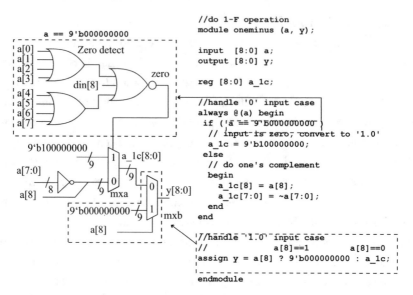

FIGURE 3.3: Implementation for 1-*F* operation.

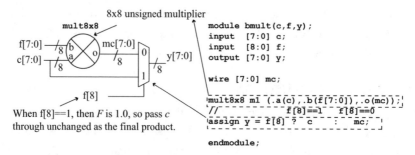

FIGURE 3.4: Multiplication of an eight-bit color operand by nine-bit blend operand.

detected by examining the most significant bit (MSb) of *F*. If *F* is not equal to zero or one, then the output is the one's complement of the lower eight-bit. The most significant bit is not included in this one's complement operation, as this would make the output value equal to one.

The multiplication operations in the blend equation have an eight-bit color operand, either C_a or C_b, and a nine-bit blend operand, either *F* or 1-*F*. When the nine-bit blend operand is not equal to one, then the multiplication result is the product of the lower eight-bit of the nine-bit blend operand and the eight-bit color operand. When the nine-bit operand is equal to one, then the product of the multiplication should be exactly equal to the eight-bit operand, which is accomplished by using a multiplexer on the output of the multiplier and testing the most significant bit of the nine-bit blend operand. The multiplication implementation is shown in Fig. 3.4; the Verilog blendmult module assumes the availability of an 8×8 multiplier component named mult8×8.

TABLE 3.2: Example blend computations

		CASE A (CNEW = CA)	CASE B (CNEW = CB)	CASE C CNEW = 0.5 * CA + 0.5*CB
F	decimal	1.0	0.0	0.5
	binary	'b100000000	'b000000000	'b010000000
1-F	decimal	0.0	1.0	0.49609375
	binary	'b000000000	'b100000000	'b001111111
Ca	decimal	0.75	0.75	0.75
	binary	'b11000000	'b11000000	'b11000000
Cb	decimal	0.25	0.25	0.25
	binary	'b01000000	'b01000000	'b01000000
Ca*F	decimal	0.75	0.0	0.375
	binary	'b11000000	'b000000000	'b01100000
Cb*(1-F)	decimal	0.0	0.25	0.12109375
	binary	'b000000000	'b01000000	'b00011111
Cnew	dec	0.75	0.25	0.49609375
	bin	'b11000000	'b01000000	'b01111111

Table 3.2 gives some example blend computations for three cases: A, B, and C. In Case A, the blend factor F is 1.0, causing C_{new} to be exactly equal to C_a. In Case B, the blend factor F is zero, causing C_{new} to be exactly equal to C_b. In Case C, the blend factor F is 0.5; note that the $1-F$ computation gives a value of 0.49609375 that is incorrect by one LSb due to the use of the one's complement to compute $1-F$. This one LSb error is propagated to the final result of 0.49609375, which should be exactly equal to 0.5 if precise arithmetic is used for the computation of 0.75 * $0.5 + (1 - 0.5)* 0.25$.

3.8   SIMPLE DATAPATHS AND THE BLEND EQUATION

Before designing an example datapath, some terms used in it are defined. The *input dataset* of a datapath contains the external values required by the datapath to perform the computation. The *output dataset* of a datapath contains the computational output of the datapath for a given input dataset. For example, the input dataset of the blend equation contains C_a, C_b, and F, while the output dataset contains C_{new}. The *latency* of a datapath measures the number of clock cycles required for a calculation on an input dataset and this number is from the first element of the input dataset to the last element of the output dataset. The *total computation time* of the datapath for an input dataset is the latency multiplied by the clock period. The *initiation period* measures how often a datapath can accept a new input dataset and is the number of clock cycles from the first element of the input dataset to the first element of the next input dataset. The *throughput* of a datapath is the number of input datasets processed per unit time; lowering the initiation period (providing input datasets more often) or decreasing the clock period increases the throughput of a datapath.

The *constraints* of a datapath determine how it is designed. Constraints are measured in both time and area (number of gates). One common constraint for datapath design is the minimum time constraint, i.e., design the datapath to perform computation in the least amount of time. Another common constraint is the minimum area constraint, i.e., design the datapath to use the minimum number of logic gates. These two constraints are contradictory to each other as performing a computation in a fewer number of clock cycles usually requires more execution units so that computations can be performed in parallel, which means more logic gates. In this chapter, we specify time constraints for a datapath as latency and initiation period values, which are measured in clock cycles. We do not specify a clock period constraint, as this is dependent upon the implementation technology such as the particular FPGA family used for the datapath.

Figure 3.5 shows the DFG of the blend equation. In a DFG, circles represent computations, with arrows linking circles to show the dataflow between computations. The operations (circles) of the DFG are labeled n1, n2, . . .nN for referral purposes. DFGs are useful in high-level synthesis tools that synthesize a datapath solution given latency and initiation period constraints. Our DFG usage is very informal and is principally used to visualize dependencies between computations; the reader is referred to [2] for a complete discussion of DFGs.

While a DFG shows the data dependencies between computations, the *datapath diagram* shows an implementation of the DFG's computation. A datapath diagram shows the computation elements and registers that are used to perform the computation and how these elements interconnect. Figure 3.6 is a datapath diagram for a naïve implementation of the blend equation. This implementation is termed naïve as it is simply a one-to-one assignment of the nodes of the DFG to execution units. This is an undesirable implementation as the execution units are *chained* together,

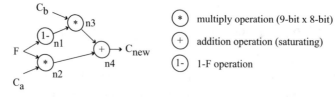

FIGURE 3.5: Dataflow graph of the blend equation.

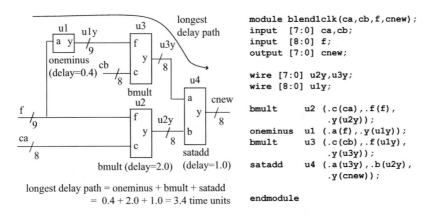

```
module blend1clk(ca,cb,f,cnew);
input   [7:0] ca,cb;
input   [8:0] f;
output [7:0] cnew;

wire [7:0] u2y,u3y;
wire [8:0] u1y;

bmult       u2 (.c(ca),.f(f),
                .y(u2y));
oneminus  u1 (.a(f),.y(u1y));
bmult       u3 (.c(cb),.f(u1y),
                .y(u3y));
satadd      u4 (.a(u3y),.b(u2y),
                .y(cnew));

endmodule
```

longest delay path = oneminus + bmult + satadd
= 0.4 + 2.0 + 1.0 = 3.4 time units

FIGURE 3.6: Naïve implementation of the blend equation.

creating a long delay path that results in a large clock period. For example purposes, relative delays of bmult = 2.0, satadd = 1.0, and oneminus = 0.4 are assumed with no time units specified. The longest combinational delay through this datapath is then $0.4 + 2.0 + 1.0 = 3.4$ time units, which forces the clock period of the system to be at least T_{cq} (register clock-to-q delay) $+ 3.4 + T_{su}$ (register setup time) assuming the inputs and outputs of the datapath are registered. Assuming that T_{cq} and T_{su} are both 0.1, this gives a system clock period of 3.6 time units.

Figure 3.7 shows a better implementation of the blend equation where DFFs have been placed after the multipliers and after the adder to break the combinational delay path, assuming that the inputs originate from a registered source. This implementation still has the 1-F calculation chained with the n3 bmult execution unit, as the 1-F operation is designed for a low combinational delay by using the one's complement operation that allows it to be chained with another execution unit. Within the datapath's Verilog code, the DFFs are implemented by the always block and are synthesized as rising edge triggered via the posedge clk in the always block's sensitivity list. Observe that the longest t_{R2R} path of Fig. 3.13 is 2.6, which is shorter than the longest combinational path of Fig. 3.11, allowing for a higher clock frequency.

The cycle-by-cycle timing for the implementation of Fig. 3.7 is shown in Fig. 3.8 for the blend computations of Table 3.2. The latency of the datapath is two clock cycles due to the two DFFs in series for any path through the datapath. The initiation period as implemented in Fig. 3.8 is two clocks as new input values are only provided every two clock cycles. Observe that this datapath takes $2 * 2.6 = 5.2$ time units to compute an output result for an input dataset, which is actually longer than the 3.6 clock period of Fig. 3.6. One reason for this is because dividing the combination delay by adding registers does not also divide the T_{cq} and setup times of the DFFs, which remain constant. Furthermore, the combinational delay path is not divided evenly when the registers are inserted. The delay of the register-to-register path that includes the adder is only 0.1 (T_{cq}) + 1.0

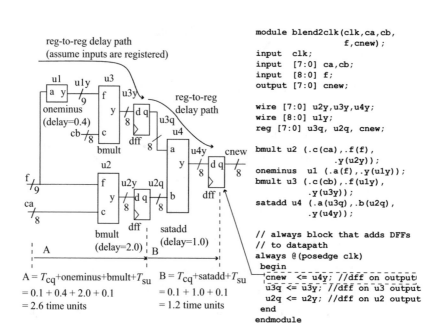

```
module blend2clk(clk,ca,cb,
                      f,cnew);
input   clk;
input   [7:0] ca,cb;
input   [8:0] f;
output  [7:0] cnew;

wire [7:0] u2y,u3y,u4y;
wire [8:0] u1y;
reg  [7:0] u3q, u2q, cnew;

bmult u2 (.c(ca),.f(f),
               .y(u2y));
oneminus u1 (.a(f),.y(u1y));
bmult u3 (.c(cb),.f(u1y),
               .y(u3y));
satadd u4 (.a(u3q),.b(u2q),
               .y(u4y));

// always block that adds DFFs
// to datapath
always @(posedge clk)
begin
  cnew <= u4y; //dff on output
  u3q <= u3y;  //dff on u3 output
  u2q <= u2y;  //dff on u2 output
end
endmodule
```

A = T_{cq}+oneminus+bmult+T_{su}
= 0.1 + 0.4 + 2.0 + 0.1
= 2.6 time units

B = T_{cq}+satadd+T_{su}
= 0.1 + 1.0 + 0.1
= 1.2 time units

FIGURE 3.7: Blend equation implementation with latency = 2.

(satadd) + 0.1 (T_{su}) = 1.2 time units, as compared to the longest path of 2.6 time units. This is not a good division of work between the datapath stages; an optimium division of labor evenly divides the delay path between the datapath stages. However, this datapath's faster clock period of 2.6 time units allows computations outside of the datapath to execute faster than that possible with the datapath of Fig. 3.6.

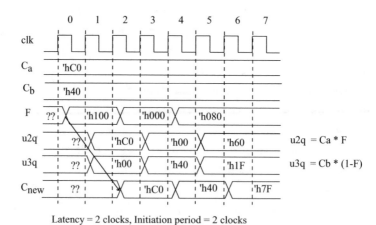

Latency = 2 clocks, Initiation period = 2 clocks

FIGURE 3.8: Cycle timing for latency = 2, initiation period = 2 clocks.

The timing diagram of Fig. 3.8 is one way to view a datapath's activity. A *scheduling table*, as shown in Table 3.3, provides another viewpoint of a datapath's activity. A scheduling table shows how DFG operations map to datapath resources such as input/output busses and execution units. Each row of the scheduling table shows the activity of the datapath resources for that clock cycle. A blank entry for a resource indicates that the resource is idle for that clock cycle. Indices such as '(0)', '(1)', etc., are used with input data values, output data values, and DFG node names to track the dataset computation that is being performed. The row entries in a schedule eventually repeat as the datapath performs the same operations on each input dataset. The last two rows in Table 3.3 form the *generalized* schedule, that is, the repeated operations on the datapath resources for each dataset. The percentage time that each datapath resource is busy during the generalized schedule is listed in the *%utilization* row of Table 3.3. Each of the datapath resources in Table 3.3 is only utilized 50% of the time as each resource is idle for one clock period of the two clock cycles that form the generalized schedule.

3.9   REGISTERING DATAPATH INPUTS VERSUS REGISTERING DATAPATH OUTPUTS

Our datapath examples place registers on the datapath outputs, and do not register the datapath inputs. The alternate choice of registering datapath inputs and leaving the outputs as unregistered is also valid, as long as consistency is followed in designing datapaths that are meant to connect together. If an external datapath with a registered output provides a value to a datapath with an unregistered input, then the communication delay from the external datapath is added to the execution unit delay that the input connects with. On large integrated circuits, the wire delay from one datapath to another can be significant if the datapaths are in different areas of the die. If the communication delay for an unregistered input value is large, then this input value should be registered in the destination datapath before being used as an input to an execution unit. The same can be said for an unregistered datapath output connected to a registered datapath input. If a datapath's input comes from off-chip or a datapath's output goes off-chip, then these signals should always be registered, as off-chip communication is slow compared to on-chip communication. Also, an unregistered datapath output should not be connected to an unregistered datapath input as the execution unit delay of the source datapath adds to the execution unit delay of the destination datapath, resulting in chained execution units.

3.10   PIPELINED COMPUTATIONS VERSUS EXECUTION UNIT PIPELINING

On viewing the datapath of Fig. 3.7 and the cycle timing of Fig. 3.8, the astute reader will realize that the datapath supports an initiation period of one clock, i.e., a new input dataset of C_a, C_b, and F can be provided for every clock. For an initiation period of one clock cycle, the second input dataset

TABLE 3.3: Schedule for latency = 2, initiation period = 2

CLOCK	RESOURCES							
	INPUT (CA)	INPUT (CB)	INPUT (F)	BMULT (U2)	ONEMINUS (U1)	BMULT (U3)	SATADD (U4)	OUTPUT (CNEW)
0	ca(0)	cb(0)	f(0)	n2(0)	n1(0)	n3(0)		
1							n4(0)	
2	ca(1)	cb(1)	f(1)	n2(1)	n1(1)	n3(1)		cnew(0)
3							n4(1)	
4	ca(2)	cb(2)	f(2)	n2(2)	n1(2)	n3(2)		cnew(1)
2i	ca(i)	cb(i)	f(i)	n2(i)	n1(i)	n3(i)		
2(i+1)							n4(i)	cnew(i-1)
%utilization	50%	50%	50%	50%	50%	50%	50%	50%

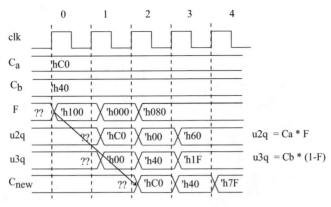

FIGURE 3.9: Cycle timing for latency $= 2$, initiation period $= 1$ clocks.

is provided before the output corresponding to first input dataset is produced. This means that the datapath has calculations on multiple datasets in progress simultaneously, with each input dataset in a different computation state. In this case, the computations for the two input datasets are said to be *pipelined*, or *overlapped*. Because the datapath resources of Fig. 3.7 are idle 50% of the time as shown by Fig. 3.8, no extra datapath resources are required to support the new initiation period of one clock cycle. Lowering the initiation period to one clock cycle doubles the throughput of the datapath, as a new result is now available with each clock instead of every two clock cycles. However, lowering the initiation period (increasing the throughput) does not affect the latency of the datapath. The cycle timing and scheduling table for an initiation period of one clock cycle are shown in Fig. 3.9 and Table 3.4, respectively. Observe that each resource is now utilized 100%, which is the best that can be achieved.

In Section 3.6, we observed that the t_{R2R} paths of Fig. 3.7 were not evenly balanced, which is undesirable as the longest t_{R2R} path determines the clock period. The excess time in the clock period for the shorter t_{R2R} paths is wasted time; distributing the delays more evenly would produce a shorter clock period. Note that the longest delay path in Fig. 3.7 contains the 1-F and multiplier units, with the multiplier having the longest delay of any execution unit. A pipeline stage inserted in the multiplier, that is, DFFs inserted within the multiplier logic, should reduce the length of this delay path. Figure 3.10 shows the blend multiplication of Fig. 3.4 modified to include a pipeline stage within the 8 × 8 multiplier. This example assumes the existence of an unsigned 8 × 8 multiplier with one pipeline stage named mult8 × 8pipe. Observe that inserting a pipeline stage in the mult8 × 8pipe component is not sufficient by itself; the other two paths through the multiplier for c[7:0] and f[8] must also have DFFs inserted so that the data streams remain synchronized when they reach the output multiplexer. If we assume that the multiplier pipeline stage perfectly divides the old combinational delay path by two, then the output and input delays of the multiplier both become

TABLE 3.4: Schedule for latency = 2, initiation period = 1

CLOCK	INPUT (CA)	INPUT (CB)	INPUT (F)	RESOURCES					
				BMULT (U2)	ONEMINUS (U1)	BMULT (U3)	SATADD (U4)	OUTPUT (CNEW)	
0	ca(0)	cb(0)	f(0)	n2(0)	n1(0)	n3(0)			
1	ca(1)	cb(1)	f(1)	n2(1)	n1(1)	n3(1)	n4(0)		
2	ca(2)	cb(2)	f(2)	n2(2)	n1(2)	n3(3)	n4(1)	cnew(0)	
3	ca(3)	cb(3)	f(3)	n2(3)	n1(3)	n3(3)	n4(2)	cnew(1)	
4	ca(4)	cb(4)	f(4)	n2(4)	n1(4)	n3(4)	n4(3)	cnew(2)	
i	ca(i)	cb(i)	f(i)	n2(i)	n1(i)	n3(i)	n4(i-1)	cnew(i-2)	
%utilization	100%	100%	100%	100%	100%	100%	100%	100%	

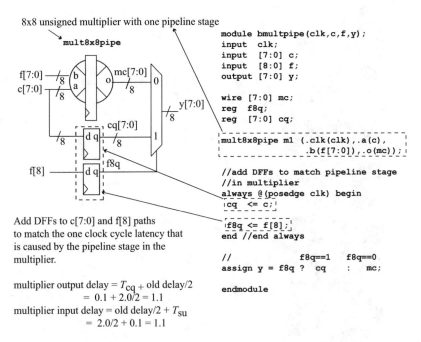

8x8 unsigned multiplier with one pipeline stage

mult8x8pipe

f[7:0]
c[7:0]

mc[7:0]

y[7:0]

cq[7:0]

f8q

f[8]

Add DFFs to c[7:0] and f[8] paths
to match the one clock cycle latency that
is caused by the pipeline stage in the
multiplier.

multiplier output delay $= T_{cq} +$ old delay/2
$= 0.1 + 2.0/2 = 1.1$
multiplier input delay $=$ old delay/2 $+ T_{su}$
$= 2.0/2 + 0.1 = 1.1$

```
module bmultpipe(clk,c,f,y);
input  clk;
input  [7:0] c;
input  [8:0] f;
output [7:0] y;

wire [7:0] mc;
reg  f8q;
reg  [7:0] cq;

mult8x8pipe m1 (.clk(clk),.a(c),
                .b(f[7:0]),.o(mc));

//add DFFs to match pipeline stage
//in multiplier
always @(posedge clk) begin
  cq <= c;

  f8q <= f[8];
end //end always

//             f8q==1   f8q==0
assign y = f8q ? cq  :  mc;

endmodule
```

FIGURE 3.10: Multiplication of an eight-bit color operand by nine-bit blend operand with pipeline stage.

equal to 1.1 time units as seen in Fig. 3.10. This decreased delay path comes at the cost of a clock cycle of latency through the blend multiplication unit.

Figure 3.11 shows the blend implementation of Fig. 3.7 modified to use the pipelined multiplier of Fig. 3.10. The longest t_{R2R} path has been reduced from 2.6 to 1.6 time units, at the cost of an extra clock cycle of latency.

The cycle timing for the blend implementation with the pipelined multiplier is shown in Fig. 3.12. The only difference between this timing and the timing in Fig. 3.9 is the extra clock cycle of latency. Table 3.5 shows the scheduling table for the blend implementation with the pipelined multiplier. The table entries for the bmultpipe units show two calculations, one for each pipeline stage of the bmultpipe unit. The extra clock cycle of latency in the bmultipipe units causes the satadd unit to remain idle until clock cycle two, as opposed to clock cycle one in the Table 3.4 schedule.

In comparing the cycle timings and schedules for the two clock cycle latency versus the three clock cycle latency solutions, a good question to ask is "When is it not advantageous to pipeline execution units?" Each clock cycle of latency is one more clock cycle that it takes for the pipeline to become full and for all execution units to become active. A pipelined datapath with a large latency is efficient as long as it has a continuous stream of input data. If the application using the datapath does

TABLE 3.5: Schedule for latency = 3, initiation period = 1

CLOCK	INPUT (CA)	INPUT (CB)	INPUT (F)	RESOURCES					
				*PIPE (U2)	ONEMINUS (U1)	*PIPE (U3)	SATADD (U4)	OUTPUT (CNEW)	
0	ca(0)	cb(0)	f(0)	n2(0)	n1(0)	n3(0)			
1	ca(1)	cb(1)	f(1)	n2(1), n2(0)	n1(1)	n3(1), n2(0)			
2	ca(2)	cb(2)	f(2)	n2(2), n2(1)	n1(2)	n3(2), n2(1)	n4(0)		
3	ca(3)	cb(3)	f(3)	n2(3), n2(2)	n1(3)	n3(3), n2(2)	n4(1)	cnew(0)	
4	ca(4)	cb(4)	f(4)	n2(4), n2(3)	n1(4)	n3(4), n2(3)	n4(2)	cnew(1)	
i	ca(i)	cb(i)	f(i)	n2(i), n2(i-1)	n1(i)	n3(i), n2(i-1)	n4(i-2)	cnew(i-3)	
%utilization 100%	100%	100%	100%	100%	100%	100%	100%	100%	

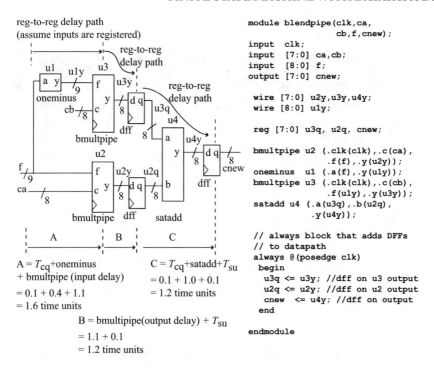

```
module blendpipe(clk,ca,
                      cb,f,cnew);
input   clk;
input   [7:0] ca,cb;
input   [8:0] f;
output  [7:0] cnew;

wire [7:0] u2y,u3y,u4y;
wire [8:0] u1y;

reg [7:0] u3q, u2q, cnew;

bmultpipe u2 (.clk(clk),.c(ca),
                  .f(f),.y(u2y));
oneminus  u1 (.a(f),.y(u1y));
bmultpipe u3 (.clk(clk),.c(cb),
                  .f(u1y),.y(u3y));
satadd u4 (.a(u3q),.b(u2q),
                  .y(u4y));

// always block that adds DFFs
// to datapath
always @(posedge clk)
  begin
    u3q <= u3y; //dff on u3 output
    u2q <= u2y; //dff on u2 output
    cnew   <= u4y; //dff on output
  end

endmodule
```

$A = T_{cq} + oneminus$
$+ bmultpipe \text{ (input delay)}$
$= 0.1 + 0.4 + 1.1$
$= 1.6 \text{ time units}$

$C = T_{cq} + satadd + T_{su}$
$= 0.1 + 1.0 + 0.1$
$= 1.2 \text{ time units}$

$B = bmultipipe \text{(output delay)} + T_{su}$
$= 1.1 + 0.1$
$= 1.2 \text{ time units}$

FIGURE 3.11: Blend equation implementation with pipelined multiplier, latency = 3.

not provide continuous input data, thus allowing the pipeline to become empty or partially empty, then the datapath throughput is significantly decreased.

Table 3.6 compares the datapaths that have been discussed to this point by clock period, latency, initiation period, and throughput.

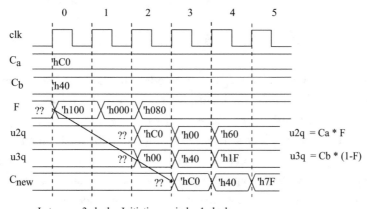

Latency = 3 clocks, Initiation period = 1 clocks

FIGURE 3.12: Cycle timing for latency = 3, initiation period = 1 clock.

TABLE 3.6: Datapath comparisons

DATAPATH	CLOCK PERIOD	LATENCY	INITIATION PERIOD	THROUGHPUT
(a) Figure 3.6	3.6	1	1	0.28
(b) Figure 3.7	2.6	2	2	0.19
(c) Figure 3.7	2.6	2	1	0.38
(d) Figure 3.11	1.6	3	1	0.63

The throughput value measures the number of input datasets processed per time unit, and is calculated by Eq. (3.2), assuming that the pipeline is filled. Decreasing either the initiation period or the clock period improves throughput, as is seen in rows (c) and (d) of Table 3.6. However, these improvements come at a cost. Decreasing the initiation period generally requires adding more datapath resources, even though this was not necessary in this simple example. Decreasing the clock period by pipelining execution units adds latency to the datapath.

$$\text{Throughput} = \frac{1}{(\text{initiation\_period} \times \text{clock\_period})} \qquad (3.2)$$

3.11   A BLEND IMPLEMENTATION WITH A SINGLE MULTIPLIER

The datapaths in Sections 3.6 and 3.7 assigned each node of the DFG of Fig 3.5 to a separate execution unit. However, in more complex datapaths, resource constraints force multiple dataflow nodes to be mapped to the same execution unit. Table 3.7 gives a schedule for a blend implementation that only contains one multiplier unit. The schedule does not use overlapped computations or pipelined execution units, and has a latency of three clocks and an initiation period of three clocks. The DFG node operations n2 and n3 are both mapped to the single multiplier unit. In this case, the execution order of n2 followed by n3 is an arbitrary choice; the execution order could be reversed.

Sharing the multiplier unit brings a new set of problems to the datapath design. The first problem is that the multiplier unit's operands now change depending upon the clock cycle. In clock

TABLE 3.7: Schedule for latency = 3, initiation period = 3, single multiplier blend implementation

CLOCK	INPUT (CA)	INPUT (CB)	INPUT (F)	RESOURCES			OUTPUT (CNEW)
				BMULT (U2)	ONEMINUS (U1)	SATADD (U4)	
0	ca(0)	cb(0)	f(0)	n2(0) ca*f→rA	n1(0)		
1				n3(0) cb*u1→rB			
2						n4(0) rA+rB→rC	cnew(0)
3	ca(1)	cb(1)	f(1)	n2(1) ca*f→rA	n1(1)		
3(i+0)	ca(i)	cb(i)	f(i)	n2(i) ca*f→rA	n1(i)		cnew(i-1)
3(i+1)				n3(i) cb*u1→rB			
3(i+2)						n4(i) rA+rB→rC	cnew(i)
%utilization	33%	33%	33%	67%	33%	33%	33%

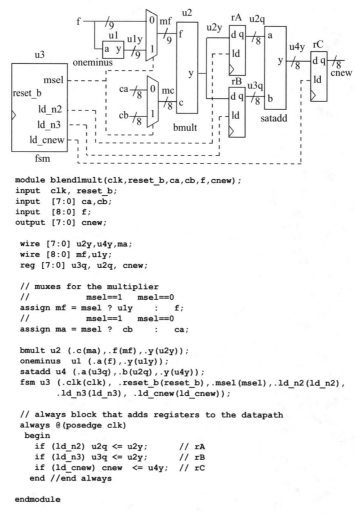

```
module blend1mult(clk,reset_b,ca,cb,f,cnew);
input   clk, reset_b;
input   [7:0] ca,cb;
input   [8:0] f;
output  [7:0] cnew;

wire [7:0] u2y,u4y,ma;
wire [8:0] mf,uly;
reg [7:0] u3q, u2q, cnew;

// muxes for the multiplier
//          msel==1    msel==0
assign mf = msel ? uly   :   f;
//          msel==1    msel==0
assign ma = msel ? cb    :   ca;

bmult u2 (.c(ma),.f(mf),.y(u2y));
oneminus  u1 (.a(f),.y(uly));
satadd u4 (.a(u3q),.b(u2q),.y(u4y));
fsm u3 (.clk(clk), .reset_b(reset_b),.msel(msel),.ld_n2(ld_n2),
        .ld_n3(ld_n3), .ld_cnew(ld_cnew));

// always block that adds registers to the datapath
always @(posedge clk)
  begin
    if (ld_n2) u2q <= u2y;      // rA
    if (ld_n3) u3q <= u2y;      // rB
    if (ld_cnew) cnew  <= u4y;  // rC
  end //end always

endmodule
```

FIGURE 3.13: Single multiplier blend implementation.

cycle $3(i + 0)$, the multiplier's operands are Ca and F, while in clock cycle $3(i + 1)$ the multiplier's operands are Cb and 1-F. This means that a multiplexer is needed on the multiplier's inputs to choose between the two sets of operands. The other problem is that a register is required to store the n2 result produced in clock cycle $3(i + 1)$ until it is needed in clock cycle $3(i + 2)$ for the n4 operation. A datapath that implements this schedule is shown in Fig. 3.13. This datapath uses *registers* instead of DFFs to break the combinational delay path and to store intermediate results. A register has a load input (LD); the register accepts a new input value only when LD is asserted and when the active clock edge occurs. By contrast, a DFF accepts a new input on each active clock edge. The three registers are named *rA*, *rB*, and *rC*. A *register transfer operation* (RTL) is added to cells in the

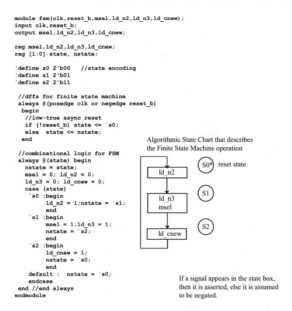

```
module fsm(clk,reset_b,msel,ld_n2,ld_n3,ld_cnew);
input clk,reset_b;
output msel,ld_n2,ld_n3,ld_cnew;

reg msel,ld_n2,ld_n3,ld_cnew;
reg [1:0] state, nstate;

`define s0 2'b00    //state encoding
`define s1 2'b01
`define s2 2'b11

//dffs for finite state machine
always @(posedge clk or negedge reset_b)
  begin
    //low-true async reset
    if (!reset_b) state <= `s0;
    else   state <= nstate;
  end

//combinational logic for FSM
always @(state) begin
  nstate = state;
  msel = 0; ld_n2 = 0;
  ld_n3 = 0; ld_cnew = 0;
  case (state)
    `s0 :begin
        ld_n2 = 1;nstate = `s1;
      end
    `s1 :begin
        msel = 1;ld_n3 = 1;
        nstate = `s2;
      end
    `s2 :begin
        ld_cnew = 1;
        nstate = `s0;
      end
    default :  nstate = `s0;
  endcase
end //end always
endmodule
```

Algorithmic State Chart that describes the Finite State Machine operation

If a signal appears in the state box, then it is asserted, else it is assumed to be negated.

FIGURE 3.14: FSM for single multiplier blend implementation.

scheduling table for each clock cycle that register writes occurs. The RTL notation "ca*f→rA" for execution unit u2 in clock zero indicates that register rA is loaded with the result of the multiplication that has the ca and f input busses as operands. Note that the rA and rB registers controlled by the ld_n2 and ld_n3 load signals have their data inputs connected to the multiplier output u2y. The ld_n2 load signal is asserted in clock cycle $3(i + 1)$ to store the n2 result, while the ld_n3 load signal is asserted in clock cycle $3(i+1)$ to store the n3 result. The ld_cnew load signal is asserted in clock cycle $3(i + 2)$ to load the output register with the satadd n4 result. The multiplexer select signal msel is negated in clock cycle $3(i + 1)$ to pass the Ca, F operands to the multiplier, while msel is asserted in clock cycle i + 1 to select $Cb, 1-F$ as the multiplier operands. As an optimization, register rB could be replaced with DFFs as its contents are only needed in the following clock cycle. The $C_{new}(i - 1)$ output value is held stable by the rC register for the duration of the computation; this might be useful if this value is used by a destination datapath. If this is not required, then register rC could also be replaced by DFFs.

A finite state machine component named FSM is responsible for driving the datapath's control lines of msel, ld_n2, ld_n3, and ld_cnew with the correct values in the appropriate clock cycles. The control signals in Fig. 3.13 are drawn with dotted lines to distinguish them from the data busses that are operated on by the execution units. The control signals and FSM component are typically not drawn in a datapath diagram; they are included here since this is the first datapath example that has required a FSM. Figure 3.14 shows the FSM implementation. Three states are required since the datapath's operation is a repeating computation covering three clock cycles.

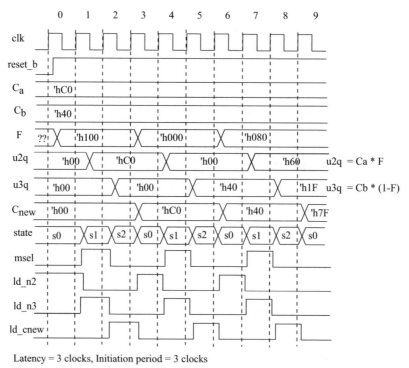

Latency = 3 clocks, Initiation period = 3 clocks

FIGURE 3.15: Cycle timing for the single multiplier blend implementation.

This FSM implementation uses two state DFFs and a grey-code encoding for the state implementation; an alternate encoding method such as one-hot encoding could have been used as well. The FSM requires an asynchronous reset input to initialize the state registers to state S0; in this example the reset signal is named reset_b and is a low-true input. The polarity choice for the reset signal, low-true or high-true, is implementation dependent.

3.12   A BLEND IMPLEMENTATION WITH HANDSHAKING

Our previous examples assumed that data is continually streaming through the datapath. However, in many cases a datapath must wait for input data to become available and must also indicate when output data is ready. Additional signals called *handshaking signals* are used by the datapath FSM for this purpose. Figure 3.16 shows the FSM of the one-multiplier blend implementation modified to add the handshaking signals irdy (input data ready) and ordy (output data ready). The differences between Fig. 3.16 and the original code in Fig. 3.14 are underlined to emphasize the changes required to support the new signals.

The FSM now remains within the S0 state until the irdy input is asserted, indicating that the input busses contain valid data, at which point the FSM transits to state S1. The ordy signal

```
module fsm(clk,reset_b,irdy,msel,ld_n2,ld_n3,ld_cnew,ordy);
input clk,reset_b,irdy;
output msel,ld_n2,ld_n3,ld_cnew,ordy;

reg msel,ld_n2,ld_n3,ld_cnew,ordy;
reg [1:0] state, nstate;

`define s0 2'b00    //state encoding
`define s1 2'b01
`define s2 2'b11

//dffs for finite state machine
always @(posedge clk or negedge reset_b)
 begin
   //low-true async reset
   if (!reset_b) begin
     state <= `s0; ordy <= 0;
   end
   else begin
     state <= nstate; ordy <= ld_cnew;
   end;
 end

//combinational logic for FSM
always @(state or irdy) begin
   nstate = state;
   msel = 0; ld_n2 = 0;
   ld_n3 = 0; ld_cnew = 0;
   case (state)
    `s0 :begin
        ld_n2 = 1;
        if (irdy) nstate = `s1;
      end
    `s1 :begin
        msel = 1;ld_n3 = 1;
        nstate = `s2;
      end
    `s2 :begin
        ld_cnew = 1;
        nstate = `s0;
      end
    default :  nstate = `s0;
   endcase
end //end always
endmodule
```

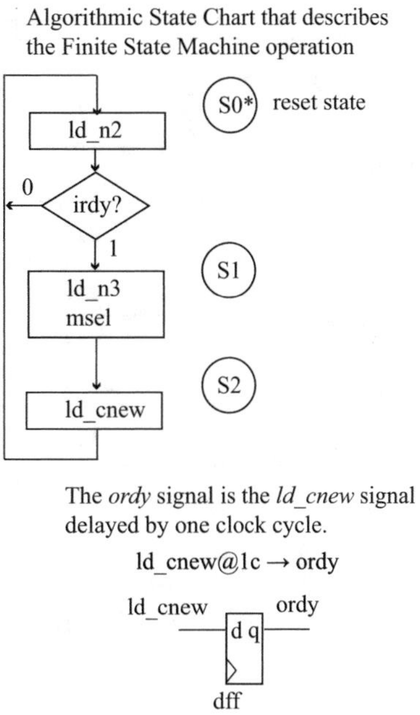

Algorithmic State Chart that describes the Finite State Machine operation

The *ordy* signal is the *ld_cnew* signal delayed by one clock cycle.

ld_cnew@1c → ordy

FIGURE 3.16: Handshaking added to FSM for single multiplier blend implementation.

is asserted for one clock cycle when valid data is placed on the *C*new output bus by delaying the ld_cnew signal that is asserted in state S2 for one clock cycle. This is implemented by a DFF that is synthesized via the Verilog assignment ordy < = ld_cnew within the always block used for the state registers of the FSM. In the Algorithm state chart (ASM chart), the ordy signal action is described by the annotation ld_cnew@1c→ordy, which reads, "ordy is assigned the value of ld_cnew, delayed by one clock". Figure 3.17 shows the cycle timing of the modified datapath for one computation; the assertion of irdy indicates valid input data and causes the computation to begin. The ordy signal is asserted when the *C*new output bus contains the computation result. The changes required to the blend1mult module of Fig. 3.13 to support the new handshaking signals are left as an exercise for the reader.

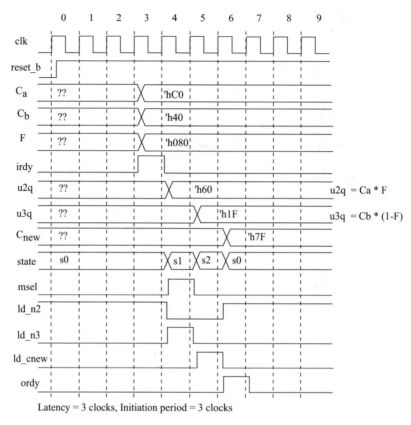

Latency = 3 clocks, Initiation period = 3 clocks

FIGURE 3.17: Cycle timing for the single multiplier blend implementation with handshaking.

3.13   A BLEND IMPLEMENTATION WITH A SHARED INPUT BUS

The previous blend implementations used separate input busses for the F, C_a, and C_b data values. However, input busses are resources in the same way as execution units are, and a designer may not have the luxury of using a separate input bus for each required input datum. External pins on an integrated circuit are extremely precious resources, and external pins are often time multiplexed between different functions. Table 3.8 gives the schedule for a blend implementation with latency = 4, initiation period = 4, uses a shared bus to input the F, C_a, and C_b data values over successive clock cycles. Only one multiplier is required; the multiplier is idle in clock cycle $i + 0$ as the C_a value is not yet available. This schedule uses a new temporary register named rF to hold the F value that is required for the n2 and n3 computations in clocks $4(i + 1)$ and $4(i + 2)$; the previous implementations assumed that the F value remained available on a separate input data bus for the duration of the computation.

TABLE 3.8: Schedule for latency = 4, initiation period = 4, shared input bus blend implementation

CLOCK	INPUT (DIN)	REGISTER (RF)	RESOURCES BMULT (U2)	ONEMINUS (U1)	SATADD (U4)	OUTPUT (CNEW)
0	f(0)	din→rF				
1	ca(0)	f(0)	n2(0) din*rF→rA	n1(0)		
2	cb(0)		n3(0) din*u1→rB			
3					n4(0) rA+rB→rC	
4	f(1)	f(1)→rF				cnew(0)
4(i+0)	f(i)	din→rF				
4(i+1)	ca(i)	f(i)	n2(i) din*rF→rA	n1(i)		cnew(i-1)
4(i+2)	cb(i)		n3(i) din*u1→rB			
4(i+3)					n4(i) rA+rB→rC	
%utilization	75%	25%	50%	25%	25%	25%

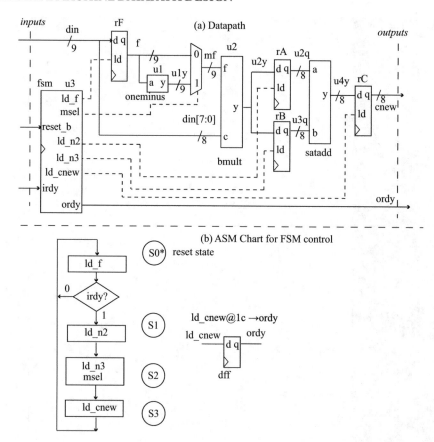

FIGURE 3.18: Shared input bus blend implementation.

Figure 3.18a shows the datapath for the blend implementation with a shared input bus. The nine-bit *din* data bus is used for the *F*, C_a, and C_b data values. The multiplexer that was used on input *c* of the *bmult* multiplier in Fig. 3.13 is no longer needed, as the C_a, C_b input values are time-multiplexed over the *din* databus.

Figure 3.18b shows the ASM chart for the datapath's FSM control; the FSM uses handshaking in the same manner as used in Fig 3.16. The Verilog code for this implementation is left as an exercise for the reader.

3.14   RECURSIVE CALCULATIONS, INITIALIZATION VERSUS COMPUTATION

The blend equation in Eq. (3.1) is a *non-recursive* equation; its output value is not dependent upon previous output values. Eq. (3.3) gives an example of a recursive equation; the Y output is dependent upon the current input value X and a previous output value Y@1. Please note that the value Y@1

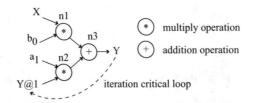

FIGURE 3.19: Dataflow graph of equation 3.3.

is the output computed from the previous input dataset, and is not the output of Y delayed by one clock cycle. A special class of digital filters known as *infinite impulse response* (IIR) filters have the general structure of (Eq. 3.3), except that multiple previous output values (Y@1, Y@2, ...Y@n) and multiple previous input values (X, X@1, X@2, ... X@k) are typically used as shown in (Eq. 3.4). The values a_i ($a_1, a_2, a_3, ... a_n$) and b_i ($b_0, b_1,..b_k$) that are multiplied by the previous output and previous input values are called the *filter coefficients*, and are determined by the filter's specifications (cutoff frequencies for low pass, band pass, high pass; roll-off constraints, etc.). Each multiplication operation is called a *filter tap*, and increasing the number of filter taps improves the filter quality.

$$Y = Y@1 \times a1 + X \times b0 \tag{3.3}$$

$$Y = (Y@1 \times a1 + Y@2 \times a2... + Y@n \times an)$$
$$+ (X \times b0 + X@1 \times b1... + X@k \times bk) \tag{3.4}$$

One of the features of a non-recursive equation is that a datapath implementation can always achieve an initiation period of one clock cycle by overlapping computations and adding the required extra resources such as input data busses, execution units, and registers. However, assuming that execution units cannot be chained, the minimum initiation period of a recursive calculation depends upon the *iteration critical loop*, which is the shortest path through the data flowgraph involving a previous output. Figure 3.19 gives the DFG of Eq. (3.3), with the iteration critical loop containing nodes n2 and n3. Each node requires one clock cycle assuming that execution unit chaining is not allowed, thus resulting in a minimum initiation period for this DFG of two clock cycles.

Table 3.9 shows a schedule for Eq. (3.3) that meets the minimum initiation period of two clock cycles. This schedule assumes that the filter coefficients are loaded into the datapath over the shared input data bus during an initialization phase, which is done before the datapath computation loop is entered.

Figure 3.20 shows the datapath and FSM control for the schedule of Table 3.17, with eight-bit data used for all calculations and 0.8 fixed-point encoding assumed. The ASM chart shows the states divided into two groups: *initialization* and *computation*. The S0 and S1 states are used to initialize the a_1, b_0 coefficient registers of the datapath with the a_1, b_0 values input over the *din* input bus in consecutive clock cycles once the *irdy* handshaking signal is asserted. States S2 and S3 form the

TABLE 3.9: Schedule for latency = 2, initiation period = 2, Eq. (3.3) implementation

CLOCK	INPUT	RESOURCES			OUTPUT
		MULT (U1)	MULT (U2)	SATADD (U3)	
0	x(0)	n1(0) b0*din→rA	n2(0) a1*rY→rB		
1				n3(0) rA+rB→rY	
2	x(1)	n1(1) b0*din→rA	n2(1) a1*rY→rB		y(0)
2(i+0)	x(i)	n1(i) b0*din→rA	n2(i) a1*rY→rA		y(i-1)
2(i+1)				n3(i) rA+rB→rY	
%utilization	50%	50%	50%	50%	50%

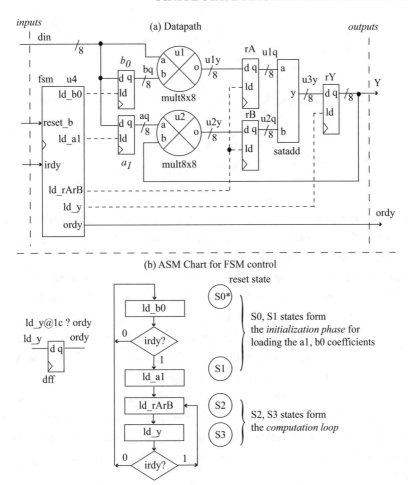

FIGURE 3.20: Datapath, FSM for equation 3.3 implementation.

computation loop, with new X values available over the *din* input bus as long as the *irdy* handshaking signal is asserted. The computation loop is exited when the *irdy* handshaking signal is negated. The *ordy* output handshaking signal is produced by delaying the *ld_y* signal of the FSM by one clock cycle. The Verilog code for this implementation is left as an exercise for the reader.

3.15   A DESIGN METHODOLOGY FOR HIGHER COMPLEXITY DATAPATHS

The previous datapath examples contained a relatively low number of operations, and scheduling the DFG operations on execution units and storing temporary results within registers was relatively straightforward. However, scheduling becomes more difficult as the target equation complexity increases, i.e., the number of operations in the target equation increases. In this section, a scheduling

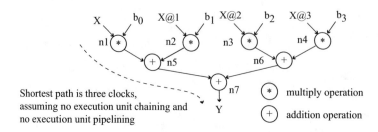

FIGURE 3.21: Dataflow graph of equation 3.5.

methodology appropriate for higher complexity datapaths is developed. This methodology does not attempt to include all of the optimizations found in behavioral synthesis methodologies [3], but rather serves to illustrate the key problems in datapath scheduling.

Equation (3.5) is a four-tap *finite impulse response* (FIR) filter, and is used as the target equation for the datapath implementations that follow. A FIR filter differs from an IIR filter (Eq. 3.4) in that it is a non-recursive equation — the filter does not use past output values. A FIR filter generally requires more filter taps than an IIR filter to achieve the same filter quality. As with the IIR equation, X@1 means the X input from the previous input dataset, and is not the X input delayed by a clock cycle. Please note that because of the regular structure of the FIR equation, an efficient datapath implementation can be done for the case of initiation period = 1, where each addition and multiplication operation are mapped to individual execution units. This equation is used in this section to illustrate the more difficult problem of mapping multiple flowgraph operations onto the same execution unit, when resource constraints prevent one-to-one mappings of operations to execution units.

$$Y = X \times b0 + X@1 \times b1 + X@2 \times b2 + X@3 \times b3 \qquad (3.5)$$

Figure 3.21 shows the DFG for Eq. (3.5). The shortest path through this DFG is three clock cycles, assuming no execution unit chaining and non-pipelined execution units. This shortest path of three clock cycles is the minimum achievable latency for this equation.

Table 3.10 shows the steps in the datapath design methodology that is followed in this section. This methodology's goal is a datapath that uses the minimum number of execution units to meet a set of target constraints.

The target constraints in this methodology is initiation period and latency, both measured in clock cycles. Step #2 computes a lower bound for each type of resource required to meet the target constraints, using Eq. (3.6). The result of Eq. (3.6) is a *lower bound* for the resource, which means that it cannot be done with any fewer resources than this value, and may actually require more than

TABLE 3.10: Datapath design methodology

STEP	ACTION
1.	Set target constraints (initiation period, latency)
2.	Compute a lower bound on the resources needed for the target constraints
3.	Attempt scheduling using the number of resources computed in step 2. If scheduling fails, go to step 4; if scheduling succeeds then go to step 5
4.	Either increase the resource(s) that has caused scheduling to fail, and loop back to step 3, or relax the constraints, and go back to step 2.
5.	Execution unit scheduling has succeeded; do register scheduling
6.	Implement the datapath

this number of resources.

$$\text{\# of resources} = \left\lceil \frac{\text{\#operations}}{\text{\#InitiationPeriod}} \right\rceil \tag{3.6}$$

For example, assume that the target constraint for a datapath implementation of Fig 3.21 is an initiation period of three clocks, and a latency of three clocks. The number of operations for a particular resource is determined by simply counting the addition or multiplication nodes in Fig. 3.21. A lower bound on the number of adders, multipliers, and input busses needed for these target constraints are given in Eqs (3.7) –(3.9). The input bus calculation of Eq. (3.8) is somewhat superfluous as the FIR equation only requires one new X input value each clock assuming that coefficient values are loading during an initialization phase, but this calculation is included to emphasize that input busses are also resources.

$$\text{\# of multipliers} = \left\lceil \frac{4}{3} \right\rceil = 2 \tag{3.7}$$

$$\text{\# of adders} = \left\lceil \frac{3}{3} \right\rceil = 1 \tag{3.8}$$

$$\text{\# of input busses} = \left\lceil \frac{1}{3} \right\rceil = 1 \tag{3.9}$$

Table 3.11 shows a scheduling attempt of Fig. 3.21 using two multipliers and one adder to meet the target constraints of latency = 3 clocks and initiation period = 3 clocks. The scheduling fails as the n7 node computation is not scheduled. In order to perform the n7 computation in clock #2, the n5 and n6 computations must both be performed in clock #1, which requires that the number

TABLE 3.11: Schedule for Figure 3.40 using two multipliers, one adder for target latency = 3, target initiation period =3

CLOCK		RESOURCES			
	INPUT	MULT (U1)	MULT (U2)	SATADD (U3)	OUTPUT
0	x(0)	n3(0)	n4(0)		
1		n1(0)	n2(0)	n6(0)	
2				n5(0)	
	Scheduling Fails! Operation n7 is not scheduled within target latency.				

of adders must be increased from one to two. However, performing the n5 and n6 computations in clock #1, requires that the n3, n4 multiply operations be performed by clock #0, which requires that the number of multipliers be increased from two to four.

Table 3.12 shows that the scheduling now succeeds with the increased resources of four multipliers and two adders for the target latency of three clocks. However, meeting this target required a doubling of the resources from their lower bound computations, which may not be acceptable if resources are limited. Relaxing the target constraints must be done if the resource requirements are too high.

If the target constraints are relaxed to initiation period = 4 clocks and latency = 4 clocks, then the new lower bound computations are shown in Eqs (3.10) and (3.11) (the input bus resource is omitted for brevity as it clearly does not affect the scheduling).

$$\text{\# of multipliers} = \left\lceil \frac{4}{4} \right\rceil = 1 \tag{3.10}$$

$$\text{\# of adders} = \left\lceil \frac{3}{4} \right\rceil = 1 \tag{3.11}$$

Table 3.13 shows that the scheduling attempt fails for these resource lower bounds, because the addition operations n5 and n7 cannot be scheduled within the target latency of four clocks. The three addition operations must begin in clock #1 if they are to be completed within the four clock latency using only one adder. If the n6 addition operation is scheduled in clock#1, then the n3 and n4 multiply operations must be scheduled in clock#0, which requires two multipliers.

TABLE 3.12: Schedule for Figure 3.40 using four multipliers, two adders for target latency = 3, target initiation period = 3

CLOCK	INPUT	MULT (U1)	MULT (U2)	MULT (U3)	MULT (U4)	SATADD (U6)	SATADD (U7)	OUTPUT
				RESOURCES				
0	x(0)	n3(0)	n4(0)	n1(0)	n2(0)			
1						n6(0) n7(0)	n5(0)	
2								
3(i+0)	x(i)	n3(i)	n4(i)	n1(i)	n2(i)			
3(i+1)						n6(i) n7(i)	n5(i)	
3(i+2)								y(i-1)
%utilization	33%	33%	33%	33%	33%	67%	33%	33%

TABLE 3.13: Schedule for Figure 3.40 using one multiplier, one adder for target latency = 4, target initiation period = 4

CLOCK	RESOURCES			
	INPUT	MULT (U1)	SATADD (U2)	OUTPUT
0	x(0)	n4(0)		
1		n3(0)		
2		n2(0)	n6(0)	
3		n1(0)		
	Scheduling fails, operations n5, n7 are not scheduled within target latency.			

Table 3.14 shows that scheduling is successful for the target latency of four clocks after the number of multipliers is increased from one to two. Assuming that this resource increase is acceptable, the datapath design can continue with register scheduling.

3.16   REGISTER SCHEDULING

Register scheduling determines how temporary results are stored in registers. This can be a complex problem if a minimum number of registers are desired as the execution unit schedule also affects the register count; fortunately registers are relatively inexpensive in terms of gate count. There may be good reasons for not using the minimum number of registers; for example, it may be desirable for the register containing a previous output result to keep this value stable throughout the computation of the new result in case it is being used by a downstream datapath. Also, using the minimum number of registers may increase the multiplexer depth in front of registers, thus creating longer t_{R2R} paths. Our register scheduling methodology only determines the registers needed for a particular execution unit schedule, and does not attempt to modify the execution unit schedule to reduce the register count.

Our register scheduling methodology begins by examining the register storage requirements of each clock as shown in Table 3.15. The *Initial* column lists the data values that are present within the datapath at the beginning of the clock cycle. The *Produced* column lists the data values that are

TABLE 3.14: Schedule for Figure 3.40 using Two Multipliers, One Adder for Target latency = 4, target initiation period = 4

CLOCK	RESOURCES				
	INPUT	MULT (U1)	MULT (U2)	SATADD (U3)	OUTPUT
0	x(0)	n3(0)	n4(0)		
1		n1(0)	n2(0)	n6(0)	
2				n5(0)	
3				n7(0)	
4($i+0$)	x(i)	n3(i)	n4(i)		y(i-1)
4($i+1$)		n1(i)	n2(i)	n6(i)	
4($i+2$)				n5(i)	
4($i+3$)				n7(i)	
%utilization	25%	50%	50%	75%	25%

either produced by computations or input to the datapath during the cycle and saved for a future clock cycle. For example, in clock cycle $i + 0$, the x value in the *Produced* column is input by the datapath during that cycle and must be saved as it becomes the *x@1* value in the next dataset computation. The *Consumed* column lists items from the *Initial* column that are no longer needed after this clock cycle. The *Total Registers* column is the total number of registers needed during that clock cycle, and is computed as *Initial + Produced – Consumed*, as registers whose values are consumed can now be used to store new values. The maximum register count in the *Total Registers* column is the number of registers required by the datapath for this schedule; in this case it is seven registers. This does not include the registers required for coefficients b_0, b_1, b_2, and b_3 as they are loaded during the initialization phase and do not change during the computation loop. The total number of datapath registers is 11 (7 + 4) once the coefficient registers are included. Observe that the scheduling of node operations in Table 3.14 affects the number of registers required for a particular clock cycle. For example, if node operations *n1*, n2 were scheduled in clock $i + 0$ instead of nodes *n3*, *n4*, then the *x@3* value would not be consumed in clock cycle $i + 0$, and the register count for that clock cycle would be seven. This does not increase the maximum number of registers for this datapath, but this may not be true for other datapaths.

TABLE 3.15: Number of required registers by clock cycle

| CLOCK | REGISTER REQUIREMENTS | | | TOTAL REGISTERS COLUMNS(1 +2 -3) |
	(1) INITIAL	(2) PRODUCED	(3) CONSUMED	
4(i+0)	x@1, x@2, x@3, y(i-1)	x, n3, n4	x@3	6
4(i+1)	x, x@1, x@2, n3, n4, y(i-1)	n1, n2, n6	n3, n4	7 (max value)
4(i+2)	x, x@1, x@2, n1, n2, n6, y(i-1)	n5	n1, n2	6
4(i+3)	x, x@1, x@2, n5, n6, y(i-1)	y(i)	n5, n6, y(i-1)	4

The registering requirements of Table 3.15 can be mapped to specific registers on a clock-by-clock basis as shown in Table 3.16. The seven registers identified in Table 3.15 are named *rA*, *rB*, *rC*, *rD*, *rE*, *rF*, and *rY*, with the register contents corresponding to the *Initial* and *Produced* columns of Table 3.15. If a register's content is changed during a clock cycle, then this is indicated by a register write operation such as "n3→rD" (the result of operation n3 is written to register *rD*) or "*rE→rA*" (the contents of register *rE* is written to register *rA*). This write operation is shown because this translates into a load line assertion for this register in the finite state machine control of the datapath. If a register's contents is no longer required after a clock cycle, then that table cell is shown as blank even though the register's contents has not physically changed (i.e., the n6 computation result in register *rF* is consumed in clock $i + 3$ and no new value is written to register *rF*, so the table cell entry for *rF* is blank in clock $i + 3$ even though the n6 computation result is still physically present). The *initial* row shows the assumed register contents at the beginning of the $i + 0$ clock cycle; the assignments of *x@1*, *x@2*, *x@3* to registers *rA*, *rB*, *rC* is an arbitrary choice. Observe that register transfers in clock $i + 3$ such as "*rE→rA*" that writes the current *x* value to *rA* is done to get ready for the next set of computations, as *x* becomes *x@1*, *x@1* becomes *x@2*, and *x@2* becomes *x@3*. The register choices made in Table 3.16 affects the multiplexing requirements of the datapath; in this methodology we do not attempt to optimize the register assignments in order to reduce the multiplexing.

The execution unit scheduling of Table 3.14 and the register content scheduling in Table 3.16 is now combined into one table that completely specifies the datapath operation, as shown in Table 3.17. The execution unit operations are now specified as RTL, such as "*rC*b3→rC*" for the n4 computation done in clock $i + 0$. The table also contains a column that contains register to register transfers such as "*rE→rA*". Observe that the choice of a particular unused register for storing a result affects the multiplexing needed for a register input. For example, the n1 and n3 computations are both written to register *rC*, while n2 and n4 are written to register *rD*. From Table 3.17, it is seen that register *rD* receives results only from multiplier unit *u2*, and thus does not require a multiplexer on its input. However, in clock cycle $i + 0$ if register *rD* had been chosen for computation n3, and *rC* for computation n4, then register *rD* would receive results from both the *u1* and *u2* multiplier units, requiring a multiplexer on the *rD* register input. After creating initial versions of the execution unit scheduling, register scheduling, and combined execution unit/register scheduling tables, the multiplexing requirements become visible and changes can be made to register assignments to reduce the number of multiplexors in the datapaths. It should be noted that high-level synthesis tools exist that perform these optimizations automatically.

The datapath and FSM implementation of the scheduling in Table 3.17 is shown in Fig. 3.22. The FSM control such as register load signals and multiplexer select signals are not shown in the datapath; the presence of these signals is assumed. Datapath diagrams such as Fig. 3.22a quickly become unwieldy as the datapath complexity increases and are also not strictly necessary, as the

TABLE 3.16: Register contents by clock cycle

CLOCK	REGISTER CONTENTS						
	RA	RB	RC	RD	RE	RF	RY
initial	x@1	x@2	x@3				y(i-1)
4(*i*+0)	x@1	x@2	n3 (n3→rC)	n4 (n4→rD)	x (x→rE)		y(i-1)
4(*i*+1)	x@1	x@2	n1 (n1→rC)	n2 (n2→rD)	x		y(i-1)
4(*i*+2)	x@1	x@2	n5 (n5→rC)		x	n6 (n6→rF)	y(i-1)
4(*i*+3)	x (rE→rA)	x@1 (rA→rB)	x@2 (rB→rC)			n6	y (n7→rY)

TABLE 3.17: Combined execution unit and register scheduling

CLOCK	DATAPATH OPERATIONS					
	INPUT	MULT (U1)	MULT (U2)	SATADD (U3)	OUTPUT	REGISTER TRANSFERS
$4(i+0)$	$x(i)$	n3(i) $rB * b2 \rightarrow rC$	n4(i) $rC * b3 \rightarrow rD$		$y(i-1)$	$x \rightarrow rE$
$4(i+1)$		n1(i) $rE * b0 \rightarrow rC$	n2(i) $rA * b1 \rightarrow rD$	n6(i) $rD + rC \rightarrow rF$		
$4(i+2)$				n5(i) $rD + rC \rightarrow rC$		
$4(i+3)$				n7(i) $rF + rC \rightarrow rY$		$rE \rightarrow rA$ $rA \rightarrow rB$ $rB \rightarrow rC$

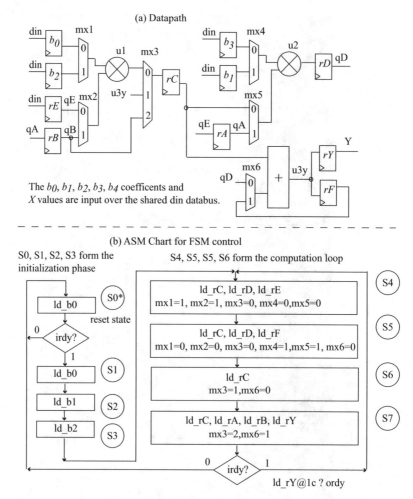

FIGURE 3.22: Datapath, FSM for implementation using Table 3.7 scheduling.

scheduling operations in Table 3.17 specify datapath operations. The Verilog code that implements the datapath is the final representation of the datapath operation, with datapath diagrams only used as an aid for visualizing the components and their interconnection that comprise the datapath. The FSM control is comprised of eight states; four states for the initialization of the coefficient registers, and four states for the compute loop. The assignments of registers to the mx1, mx2 and mx4, mx5 multiplexer inputs were done so that the select lines of these two pairs of multiplexors can be connected together. Thus, the number of multiplexer select signals in the ASM chart can be reduced from what is shown, as the mx1, mx2 and mx4, mx5 signals have the same values in each state and thus each pair can be driven by one signal. The assignments of inputs to the mx3 and mx6 multiplexer were arbitrarily chosen.

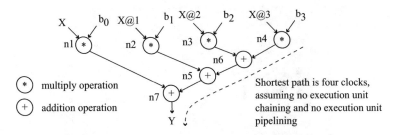

FIGURE 3.23: Restructured flowgraph for equation 3.5.

3.17   FLOWGRAPH TRANSFORMATIONS, OVERLAPPED COMPUTATIONS REVISITED

In the previous section, two multipliers were required for a latency $= 4$, initiation period $= 4$ solution to the flowgraph of Fig. 3.21.

Table 3.18 shows an attempt to remove one of the multipliers by increasing the target latency from four clocks to five clocks. However, the schedule fails because the last addition operation, n7, is not scheduled within the target latency.

For the scheduling to succeed with a latency of five clocks and only one multiplier, the three multiplication operations have to begin in clock cycle #2, with one multiplication done per clock cycle. Fortunately, the multiply-accumulate operations in Eq. (3.5) are associative, allowing the flowgraph to be *restructured* as shown in Fig. 3.23.

This illustrates the dependency of scheduling on flowgraph structure; automated high-level synthesis tools will restructure a flowgraph when searching for a scheduling solution that meets target latency and target initiation period constraints.

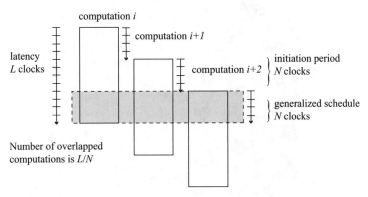

FIGURE 3.24: Overlapped computations.

TABLE 3.18: Schedule for Figure 3.40 using one multiplier, one adder for target latency = 5, target initiation period = 5

| CLOCK | INPUT | RESOURCES | | OUTPUT |
		MULT (U1)	SATADD (U2)	
5(i+0)	x(i)	n4(i)		
5(i+1)		n3(i)		
5(i+2)		n2(i)	n6(i)	
5(i+3)		n1(i)	n5(i)	
5(i+4)				

Scheduling fails, operation n7 is not scheduled within target latency.

TABLE 3.19: Schedule for Figure 3.44 using one multiplier, one adder for target latency = 5, target initiation period = 5

CLOCK	RESOURCES			
	INPUT	MULT (U1)	SATADD (U2)	OUTPUT
5(i+0)	x(i)	n4(i)		y(i-1)
5(i+1)		n3(i)		
5(i+2)		n2(i)	n6(i)	
5(i+3)		n1(i)	n5(i)	
5(i+4)			n7(i)	
%utilization	20%	80%	60%	20%

3.18   OVERLAPPED COMPUTATIONS REVISITED

As stated earlier, overlapping computations for input datasets increases throughput usually at the cost of additional resources. The methodology of Table 3.10 can also be used for overlapped computations. When determining the initiation period (N) and latency (L) constraints for overlapped computations, the initiation period should be evenly divisible into the latency. The latency divided by the initiation period (L/N) is the number of *overlapped computations* in the design, and the generalized schedule is equal to the initiation period of N clocks.

As an example, choose a target latency of four clocks, and a target initiation period of two clocks for the flowgraph of Fig. 3.21. Using Eq. (3.6), the lower bounds on the multiplier and adder execution units are given in Eqs (3.12) and (3.13).

$$\text{\# of multipliers} = \left\lceil \frac{4}{2} \right\rceil = 2 \tag{3.12}$$

$$\text{\# of adders} = \left\lceil \frac{3}{2} \right\rceil = 2 \tag{3.13}$$

The number of overlapped computations is 4/2 = 2, so the generalized schedule contains computations for datasets i and $i-1$, and the output value for computation $i-2$. Table 3.20 shows that scheduling succeeds for latency = 4 and initiation period = 2 using the lower bound estimates for execution units. Observe that the operations mapped to execution units are chosen such as to repeat the same operations on the execution unit for the initiation period's two clocks in order to reach a

TABLE 3.20: Schedule for Fig. 3.21 using two multipliers, two adders for target latency $= 4$, target initiation period $=2$

CLOCKS	INPUT	RESOURCES				OUTPUT
		MULT (U1)	MULT (U2)	SATADD (U3)	SATADD (U4)	
0	x(0)	n3(0)	n4(0)			
1		n1(0)	n2(0)	n6(0)		
2	x(1)	n3(1)	n4(1)	n5(0)		
3		n1(1)	n2(1)	n6(1)	n7(0)	
4	x(2)	n3(2)	n4(2)	n5(1)		y(0)
5		n1(2)	n2(2)	n6(2)	n7(1)	
$2(i+0)$	x(i)	n3(i)	n4(i)	n5(i-1)		
$2(i+1)$		n1(i)	n2(i)	n6(i)	n7(i-1)	y(i-2)
%utilization	50%	100%	100%	100%	50%	50%

generalized schedule. For example, it would not work to schedule the n6, n5, and n7 operations all on the u3 adder as this cannot be repeated within the two clocks of the initiation period.

Table 3.21 show that the temporary registers required by this schedule is eight, so the total number of registers needed for the datapath, including the four coefficient registers, is 12. Assuming the clock period remains the same, doubling the throughput has only cost one additional register and one extra adder. The reason for this small increase in resources is because of the low *%utilization* of the resources in the latency $= 4$, initiation period $= 4$ solution of Table 3.14.

The remaining detailed register scheduling and datapath design is left as an exercise for the reader.

3.19   SUMMARY

DFGs are useful tools for visualizing the data dependencies of a computation. Latency and initiation period constraints determine the number of registers and execution units required to implement a particular computation. A scheduling table is used to map computations to available execution units and registers. Overlapped computations and pipelined executions are both useful techniques for increasing the throughput of a datapath.

3.20   SAMPLE EXERCISES

1. Create a Verilog implementation of the datapath in Fig. 3.18.

2. Create a Verilog implementation of the datapath in Fig. 3.20.

3. Design a datapath with latency $= 3$, and initiation period $= 3$ for the DFG of Fig. 3.19 using multiplier units with one pipeline stage; use the minimum number of adder and multiplier units that meets these constraints.

4. Modify the schedule of Table 3.8 for latency $= 4$, initiation period $= 2$ and do a Verilog implementation of the datapath.

5. Do a Verilog implementation of the datapath in Fig. 3.22.

6. Use Table 3.20 and Table 3.21 to complete a schedule that contains all of register transfer operations for this datapath, create the ASM chart for the required FSM control, and implement the datapath in Verilog.

Equation 3.14 implements an operation known as *bilinear filtering* in which a new color C_{new} is produced from four colors C_{00}, C_{01}, C_{10}, C_{11} using two blend factors, u and v. As an example, if $v = 0.5$ and $u = 0.5$, then C_{new} is an equal blend of each color ($C_{new} = 0.25^*C_{00} + 0.25^*C_{01} + 0.25^*C_{10} + 0.25^*C_{11}$). The data types and operations in Eq. 3.14 are the same as in the blend equation. The colors are 0.8 fixed-point values, while u, v are nine-bit values encoded in the same manner as F in

TABLE 3.21: Number of required registers by clock cycle

CLOCK	REGISTER REQUIREMENTS			
	(1) INITIAL	(2) PRODUCED	(3) CONSUMED	TOTAL REGISTERS (1+2-3)
2(*i*+0)	x@1, x@2, x@3, n1(i-1), n2(i-1), n6(i-1), y(i-2)	x, n3(i), n4(i),n5(i-1)	x@3,n1(i-1),n2(i-1)	7+4-3 = 8 max value
2(*i*+1)	x, x@1, x@2, n3(i), n4(i), n5(i-1), n6(i-1),y(i-2)	n1(i), n2(i), n6(i),y(i-1)	n3(i),n4(i),n5(i-1), n6(i-1),y(i-2)	8+4-5 = 7

the blend equation.

$$C_{new} = C00 \times (1 - v) \times (1 - u) + C01 \times (1 - v) \times u$$
$$+ C10 \times v \times (1 - u) + C11 \times v \times u \qquad (3.14)$$

Figure 3.25 shows a DFG for Eq. (3.14) that assumes a single nine-bit input databus, with the u, v blend factors input during the datapath initialization phase and multiple four-tuples of C_{00}, C_{01}, C_{10}, and C_{11} input during the computation loop for use with these blend factors. The square boxes around C_{01}, C_{10}, and C_{11} and the arrows linking C_{00}, C_{01}, C_{10}, and C_{11} indicate that these are input operations over a shared input bus.

The following questions reference Eq. (3.14) and Fig. 3.25. Use the minimum number of execution units in all implementations.

7. Using the methodology of Table 3.10, design a datapath that has latency $= 6$ clocks and initiation period $= 6$ clocks. Assume that C_{00}, C_{01}, C_{10}, and C_{11} are available in successive clock cycles in the first four clocks of the initiation period.

8. Using the methodology of Table 3.10, design a datapath that has latency $= 8$ clocks and initiation period $= 4$ clocks. Assume that C_{00}, C_{01}, C_{10}, and C_{11} are available in successive clock cycles in the four clocks that comprise the initiation period.

9. If multiplier units with one pipeline stage are used in, then the shortest path becomes eight clocks. Using multiplier units with one pipeline stage, design a datapath that has latency $=$ eight clocks and initiation period $=$ eight clocks.

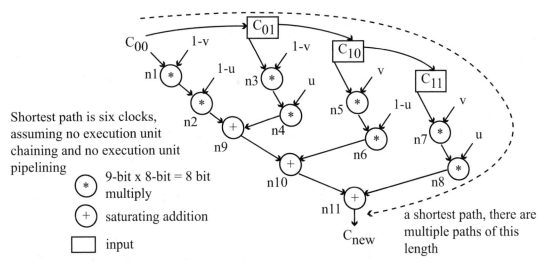

FIGURE 3.25: Dataflow Graph for Equation 3.14

10. If multiplier units with one pipeline stage are used in, then the shortest path becomes eight clocks. Using multiplier units with one pipeline stage, design a datapath that has latency = eight clocks and initiation period = four clocks.

APPENDIX: IS DATAPATH SCHEDULING A VALID TOPIC FOR MODERN DIGITAL SYSTEM DESIGN?

This chapter discusses datapath scheduling at length; it may be argued that with the high gate counts of modern FPGAs, the need for resource sharing has passed and that modern designs are mostly done in a parallel, fully pipelined manner to emphasize throughput. Another argument can be made that individual multipliers and adders are passé when tools like Xilinx Coregen can automatically generate a 1024-Point Complex Fast Fourier Transform block or the AccelDSP tool can generate a Verilog implementation for an arbitrary Matlab function.

It is the author's contention that a fundamental grounding in the concepts of latency and throughput in relation to computational intensive FSM/datapaths is important, even in the context of modern FPGAs that can contain hundreds of thousands to millions of gates. At some point, a designer will be concerned with latency/throughput of a design, and the gate count tradeoffs associated with latency/throughput. A modern designer may be using double-precision floating-point units as executions units instead of fixed-point adders and multipliers, but the finite state machine task of sequencing operations on those units and storing intermediate results will remain unchanged. Furthermore, a modern designer is usually part of a team, and may be given the task of generating a computation block to be used in a much larger design, and will probably be given latency, throughput, and clock speed constraints on that design.

Finally, even if a modern designer has a high level synthesis tool that can automatically generate an RTL design from a high-level language description in C (or some other programming language), it is important that the designer has a firm understanding of latency/throughput and clock speed because they will almost certainly be the constraints given to the high level synthesis tool when generating the design.

Datapath scheduling/RTL coding versus high-level synthesis and pre-generated IP blocks can be likened to programming in assembly language versus programming in a high level language. A modern programmer may never have the need to program in assembly language. However, it can be assured that the programmer has training in assembly language in order to understand the linkage between a high level language (HLL) and the target assembly language, and to understand the role that a compiler plays in the transformation of HLL code to assembly, and effect of HLL data types and compiler code optimizations on resulting code size and execution speed. Also, if nobody understands assembly language (datapath scheduling/RTL), who will write the compilers (write high-level synthesis tools or build embedded IP blocks)?

3.21   REFERENCES

[1] Kai Hwang, Computer Arithmetic Principles, Architecture and Design, Wiley, 1979.

[2] S. S. Bhattacharya, P.K. Murthy et al., Software Synthesis from Dataflow Graphs, Kluwer Academic Publishers, 1996.

[3] Sumit Gupta, Rajesh Gupta et al. SPARK:: A Parallelizing Approach to the High-Level Synthesis of Digital Circuits, Springer 2005, pp 262.

CHAPTER 4

Embedded Memory Usage in Finite State Machine with Datapath (FSMD) Designs

This chapter explores usage of different types of embedded memories such as read-only memories (ROMs), single-port random access memories (RAMs), first-in first-out buffers (FIFOs), and dual-port RAMs in finite state machine with datapath designs.

4.1 LEARNING OBJECTIVES

After reading this chapter, you will be able to perform the following tasks:

1. Discuss the operational differences between synchronous and asynchronous embedded memories, and between single-port, dual-port, and FIFO memories.

2. Implement FSM/datapaths that incorporate single-port synchronous RAMs.

3. Discuss application scenarios for FIFOs and dual-port memories.

4. Use two-phase and four-phase handshaking for data transfer.

5. Use a two-flop synchronizer for asynchronous input synchronization.

4.2 INTRODUCTION TO EMBEDDED MEMORIES

Modern FPGAs have various types of embedded memories available for designer usage. A simple type of embedded memory block is the asynchronous $K \times N$ ROM, as shown in Fig. 4.1a. This memory is labeled as *asynchronous* because there is no clock signal for controlling access to the memory's contents. The memory is labeled as *read-only*, because its contents are fixed at FPGA configuration time; there is no method by which the application can modify the memory's contents. The $K \times N$ parameters give the memory's *organization*; the memory has K locations with each location containing N bits, thus providing a total data storage of $K \times N$ bits. An *address* bus, labeled as addr, is used to access the memory's contents; the width of the address bus is $\log_2(K)$. The output data bus, labeled as dout, carries the data of the memory location specified by the address bus. An asynchronous ROM is a combinational logic device; the output (dout) changes after some delay from an input (addr) change. This propagation delay from a change in address value to a stable data output value is the memory's *access time* (T_{ACCESS}). In general, larger embedded memories have

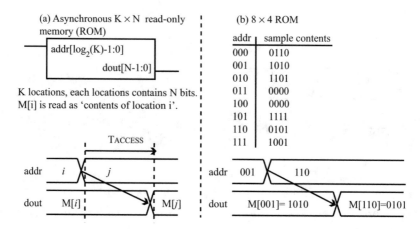

FIGURE 4.1: Asynchronous $K \times N$ read-only memory (ROM).

longer access times. Figure 4.1b shows sample contents for an 8 × 4 ROM; this memory requires a three-bit address bus ($\log_2(K)$) and a four-bit data output bus.

Figure 4.2 shows a synchronous version of a $K \times N$ ROM. DFFs are placed on the address inputs (i.e., these inputs are *registered*), thus latching the address inputs on a rising clock edge. The data output bus is available in both registered and unregistered versions. A designer might use the registered dout version if the ROM's access time is large and the designer did not want the ROM's access time summed with the datapath delay that follows the ROM's output. This is similar to the methodology used in Chapter 3 in which registers are placed between execution units (adders, multipliers) to break long combinational paths, reducing critical path length and increasing system clock frequency. The tradeoff associated with using the registered dout bus is a clock cycle of latency for data access; the registered dout value in the current clock cycle corresponds to the memory contents of the address bus value latched on the rising clock edge of the previous clock cycle. By contrast, the unregistered dout bus contains the memory contents of the address value latched on the rising clock edge of the current clock cycle. The registered dout value is available at T_{cq} propagation delay after the rising clock edge; T_{cq} is less than T_{ACCESS} time. It should be noted that the availability of both registered and unregistered dout buses in synchronous embedded memories is a design decision made by the FPGA vendor and thus will vary by FPGA vendor and by FPGA family. In this text, the assumption is made that both registered and unregistered dout buses are available.

Random Access Memory (RAM) is an embedded memory block whose contents can be modified under application control. Figure 4.3 shows an asynchronous $K \times N$ RAM; the additional signals on this embedded memory block when compared to the asynchronous ROM of Fig. 4.1 are the data input bus (din) and write enable (we) input. New data on the din bus is written to the current address location when the we enable signal experiences a high-to-low transition; there

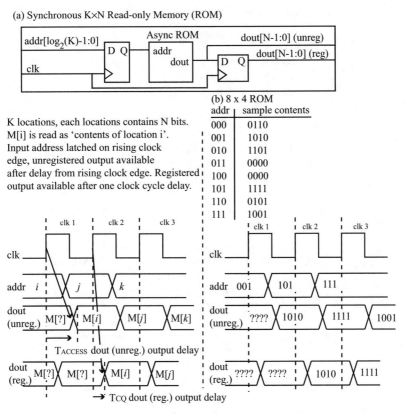

FIGURE 4.2: Synchronous $K \times N$ read-only memory (ROM).

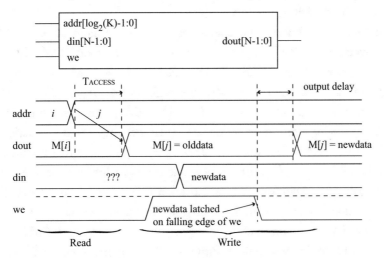

FIGURE 4.3: Asynchronous $K \times N$ random access memory (RAM).

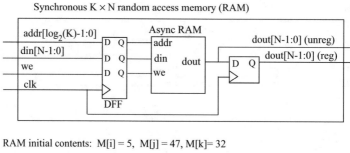

Synchronous K × N random access memory (RAM)

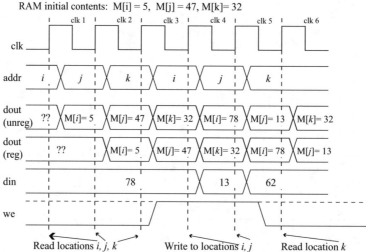

FIGURE 4.4: Synchronous $K \times N$ random access memory (RAM).

is also a minimum high pulse-width requirement on the we signal with setup (tsu) and hold (thd) constraints for din on the falling we edge.

Figure 4.4 shows read and write operations for a synchronous $K \times N$ RAM. The read operation for a synchronous RAM is the same as for a synchronous ROM, the address input is latched on the rising clock edge and output data is available either as an access time later (unregistered dout) or a T_{cq} time after the next rising clock edge (registered dout). In Fig. 4.4, clock cycles four and five demonstrate write operations to the RAM. The addr, din, and we inputs are latched on the rising clock edge; a logic one value on the we signal indicates a write operation. Location i is written with the value 78 (din bus value) in clock cycle four; observe that the unregistered dout bus reflects this new value as t_{pd} (at least T_{ACCESS} and it may be longer depending on the memory) after the rising clock edge of clock cycle four. Location j is written in clock five with the value 13. The din bus value does not affect memory operation when we is negated.

Synchronous RAMs are almost always preferred over asynchronous RAMs in designs in order to avoid problems with timing uncertainty during write operations. Figure 4.5 shows a finite state machine (FSM) connected to an asynchronous RAM, with the timing diagram illustrating a write

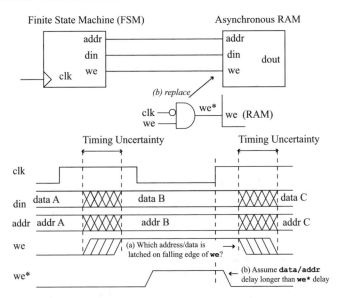

FIGURE 4.5: A problem with using an asynchronous RAM with a FSM.

operation. New `addr`, `din`, and we values are provided by the FSM some delay after the rising clock edge. This delay is dependent on how the signals are generated by the FSM (registered only, or registered plus combinational encoding) as well as wiring delays between the FSM and the RAM. Wire delay in FPGAs can be significant and can also vary significantly depending on the number of programmable switches that a signal passes through between the blocks. This timing uncertainty is problematic during a write operation as the input data and address values that are latched on the falling edge of the write enable signal are unknown. This problem is sometimes attacked by AND'ing the we signal from the FSM with the inverse of the clock signal and using this new signal (we*) as the RAM we. However, this approach relies on the assumption that the address and data input bus values have a longer delay than we*, which is an assumption that may not be true and whose timing may be violated if routing delays change between the FSM block and RAM.

The timing problems in Fig. 4.5 can be avoided by using a synchronous RAM, as shown in Fig. 4.6. The `data/addr/we` signals to the synchronous RAM only have to satisfy t_{su} and t_{hd} relative to the rising clock edge. The timing uncertainty for these signals can be an issue for t_{hd}, but t_{pd} of the `data/addr/we` signals after the rising clock edge is typically much larger than the RAM t_{hd}, which is either zero or very small. The astute reader may observe that because a synchronous RAM is an asynchronous RAM with registered inputs, the race condition between the `addr/din` signals and the we signal is simply moved inside the synchronous RAM block. This is true, but it is the responsibility of the synchronous RAM designer to solve this timing problem, and it is not an issue for a designer who wishes to use synchronous RAM blocks since correct operation is guaranteed as long as the input t_{su} and t_{hd} are met.

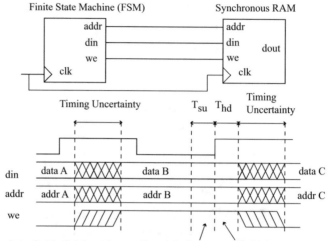

The **data/addr/we** inputs must only satisfy the setup and hold times of
the synchronous RAM; the timing uncertainty of these signals is not an issue.

FIGURE 4.6: Using a synchronous RAM with a FSM.

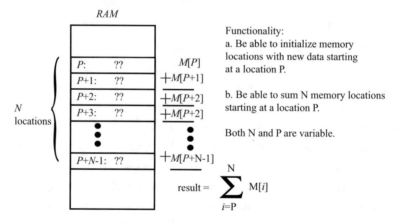

FIGURE 4.7: Memory sum overview.

4.3   SAMPLE APPLICATION: MEMORY SUM

Figure 4.7 gives an overview of a simple application used to illustrate a datapath design that contains
an embedded synchronous RAM. The datapath's functionality consists of two operation modes:

- *Initialization*: the datapath initializes the RAM's content's starting at a location P. Both P
  and initialization data are provided from an external input data bus.

- *Computation:* the datapath sums the contents of N locations, starting at location P. Both N
  and P are specified by an external input data bus, and with the result given on an external
  output data bus.

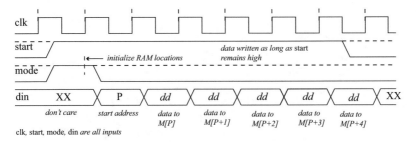

FIGURE 4.8: Initialization mode timing specification.

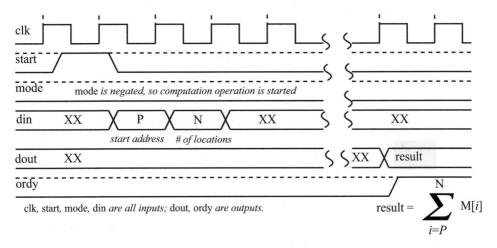

FIGURE 4.9: Computation mode timing specification.

Datapath operation is controlled by assertion of a start input, with a mode input determining if initialization or computation is performed.

The cycle timing specification for the initialization operation is shown in Fig. 4.8. The combination of start = 1 and mode = 1 causes the initialization operation to begin. The starting address P for the initialization operation is provided on the din input bus in the clock cycle following start assertion. Memory locations $M[P]$, $M[P+1]$, $M[P+2]$, etc., are written in successive clock cycles with data provided on din; locations are written as long as start is asserted (Figure 4.8 shows writes to only four locations; more locations could have been written). The negation of start signals the end of the initialization operation.

Figure 4.9 gives the computation mode timing specification. The start address (P) and number of locations to sum (N) are provided in the first two clock cycles after start assertion with mode = 0. At some later time, the output ready (ordy) output is asserted by the datapath when the result is available on the dout data bus. The number of clock cycles required for the computation is implementation dependent.

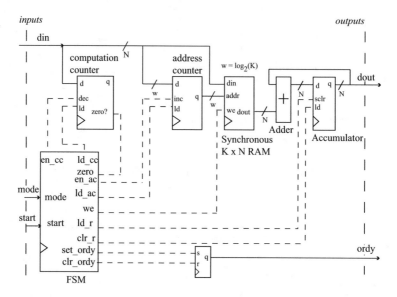

FIGURE 4.10: Memory sum datapath.

A datapath (Fig. 4.10) and finite state machine (ASM chart is shown in Fig. 4.11) performs the required operations of initialization and computation. The datapath particulars are:

- The *address* counter provides the RAM address; it is used to sequentially access memory locations during both initialization and computation operations. The counter is loaded with (P) at the start of both operation modes, and has an increment by-one functionality.

- The *computation* counter tracks the number of locations remaining to be summed during computation operation and is loaded with (N) to be summed at the beginning of this operation. The computation operation is halted when the count value reaches zero. The counter has a decrement by one functionality.

- The adder coupled with an output register provides an *accumulator* functionality, that is, successive additions add the register value with the contents of the currently accessed memory location. The register has a synchronous clear function since the register value must be zero for the first addition. The dout bus is the accumulator output.

- A synchronous $K \times N$ RAM is used as the embedded memory block.

- A set/reset flip-flop (SRFF) is used to implement the output ready (ordy) signal; an SRFF is useful when a signal must be asserted for several clock cycles.

The FSM sequences the actions on the datapath according to the ASM chart given in Fig. 4.11. State S0 waits for start assertion, and then branches to the first states of the initialization operation or computation operation based on the mode input.

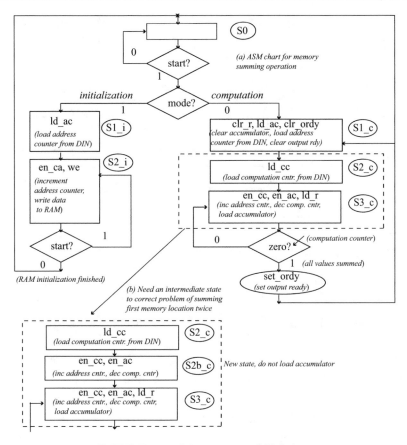

FIGURE 4.11: Memory sum ASM chart.

The initialization operation is straightforward. The first state S1_i loads the starting address into the address counter by asserting the address counter's `ld` input. The second state S2_i writes data values in the RAM by asserting the RAM's write enable; the input data is provided on the `din` data bus. The address counter is incremented in S2_i by assertion of the address counter's `inc` input. State S2_i returns to state S0 when `start` is negated. Fig. 4.12 is a timing diagram for the initialization operation with example data, and contains both external and internal signals. Data is written to locations four through eight on the leading rising edges of clocks four through eight. Observe that even though `start` is negated in clock cycle seven, the data in this clock cycle is written to RAM as specified in Fig. 4.8.

Two versions of the computation operation are provided— an incorrect version of three states (S1_c, S2_c and S3_c) and a correct version of four states (S1_c, S2_c, S2b_c, and S3_c). The incorrect version appears to be a straightforward implementation of the computation operation of Fig. 4.9 in that the starting address and locations to be summed are captured in states S1_c and S2_c, with state

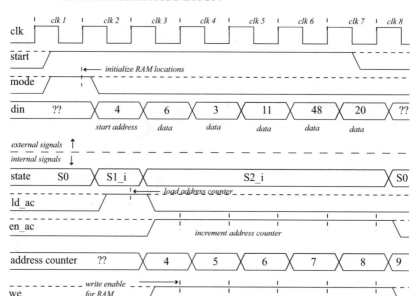

FIGURE 4.12: Initialization operation showing both external and internal signals for sample data.

S3_c is used to sum the memory contents. However, Fig. 4.13 illustrates the reason for the incorrect behavior by attempting to sum two locations, starting at location five. In the first clock cycle of state S3_c (clock 4), the memory dout bus contains $M[5] = 3$, the accumulator value is zero, and the adder output is $3 + 0 = 3$. The accumulator load signal is asserted in S3_c, so in clock five the accumulator becomes three, and the address counter is incremented to location six. However, even though the address counter value is now six, this value is not latched into the RAM until the next clock cycle, and thus the RAM dout remains at $M[5] = 3$ for clock five. This means that at the end of clock five, the new value loaded into accumulator is $3 + 3 = 6$, causing the first location to be included twice in the accumulated sum. The next clock produces $M[6] = 11 + 6$ for a final result of 17, which is incorrect. The correct result should be $M[5] + M[6] = 3 + 11 = 12$.

There are multiple ways to correct the errant behavior of Fig. 4.13; one solution is to not assert the accumulator load line in the first clock cycle after state S2_c. This is done by inserting a new state named S2b_c between states S2_c and S3_c; state S2b_c increments the address counter and decrements the computation in the same way as state S3_c, but it does not load the accumulator register. Fig. 4.14 shows the datapath/FSM operation with the new S2b_c state producing the correct sum of $M[5] + M[6] = 3 + 11 = 14$.

4.4   FIRST-IN, FIRST-OUT BUFFER

Another type of embedded memory block is a first-in, first-out (FIFO) buffer, which is a synchronous RAM block that has additional logic to give it a specialized behavior. Figure 4.15 shows the concep-

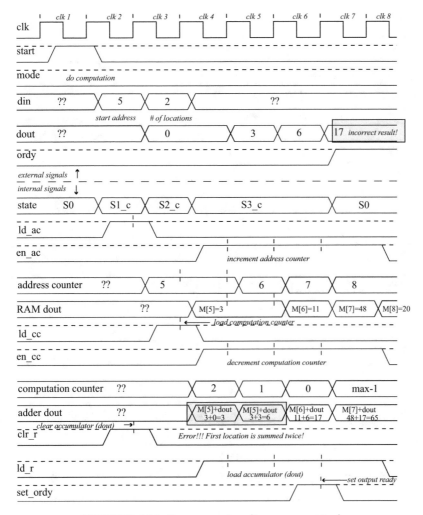

FIGURE 4.13: Sum operation (incorrect version).

tual operation for an eight-entry FIFO. A FIFO has a write port for placing data into the FIFO, and a read port for removing data from the FIFO. Figure 4.15a shows an empty eight-element FIFO. A write operation in Fig. 4.15b places *dataA* into the buffer, followed by a second write of *dataB* in Fig. 4.15c. Read operations in Fig. 4.15d and Fig. 4.15e first removes *dataA* and then *dataB*, thus illustrating FIFO nature of the buffer. Figure 4.15f shows a full FIFO after eight successive write operations.

Figure 4.16 provides two sample uses of a FIFO in a digital system. One common use is for buffering data from an external input channel as shown in Fig. 4.16a. Many input channels have the characteristic that data arrives in irregular bursts, and the individual data elements cannot always be processed by the digital system as they arrive, since the system may be busy with other tasks. The

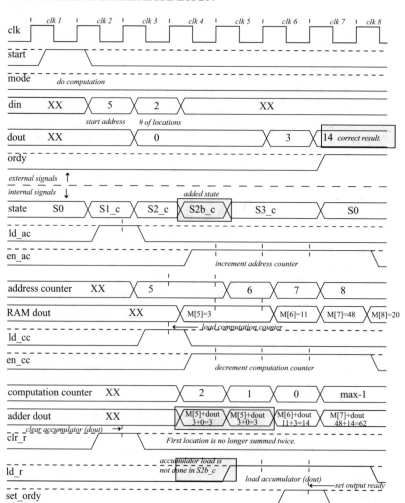

FIGURE 4.14: Sum operation (correct version).

FIFO holds the data until the system is ready for input processing. If handshaking signals are not used to regulate the data flow of the input channel, then the FIFO size is chosen to accommodate the maximum expected number of data elements to arrive between input processing tasks by the digital system.

Another typical FIFO usage is for data transfer between cooperating FSM/datapaths operating in different clock domains as shown in Fig. 4.16b. Data is written to the FIFO synchronized by clock domain A, and removed from the FIFO synchronized by clock domain B. Data transfer between two independent clock domains is an *asynchronous* transfer, that is, data can arrive at any time and is not synchronized to the receiver's active clock edge. This uncertainty in data arrival can cause

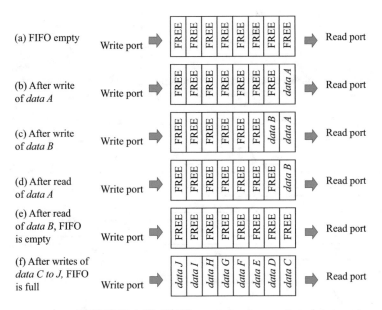

FIGURE 4.15: FIFO conceptual operation.

t_{su} and t_{hd} violations in the receiver's input register, resulting in a corrupted data transfer. A FIFO that supports independent read and write clocks is one method for solving this asynchronous data transfer problem.

The design of a FIFO with independent read/write clocks is challenging from a timing perspective, and is beyond the scope of this text, but FPGA vendors provide these as ready-to-use

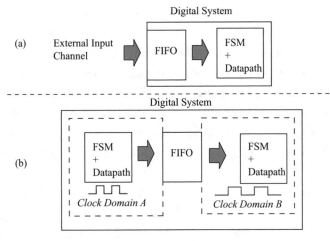

FIGURE 4.16: FIFO usage.

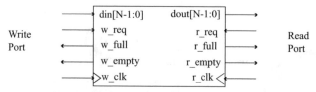

For simplicity, timing diagrams shown with common clock and common output status:
(r_clk = w_clk = clk, r_full = w_full = full, r_empty = w_empty = empty)

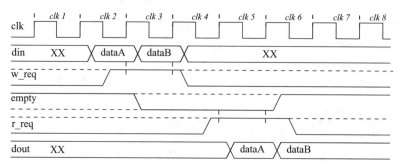

FIGURE 4.17: FIFO interface.

embedded memory blocks. Figure 4.17 shows a sample interface for a FIFO with independent read/write clocks. The write port consists of the write clock (w_clk), input data bus (din), write request input (w_req), empty status output (w_empty), and full status output (w_full). Data is written to the FIFO on the active edge of w_clk when the w_req input is asserted. The w_empty output is asserted when the FIFO is empty, and w_full is asserted when the FIFO is full, with transitions synchronized to the write clock. The read port consists of the read clock (r_clk), output data bus (dout), read request input (r_req), empty status output (r_empty), and full status output (r_full). Data is read from the FIFO on the active edge of r_clk when the r_req input is asserted. The timing diagram in Fig. 4.17 shows *dataA*, *dataB* written to an empty FIFO in clocks three and four, and data read from the FIFO in clocks five and six. Observe that the empty status output is negated after the write of *dataA* to the FIFO, and is asserted after the read of *dataB* from the FIFO. For simplicity, the timing diagrams assumes common clocks for the read and write ports. It must be noted that the timing details of FIFOs with independent read/write clocks can vary significantly from one FPGA vendor to another, and even between FPGA families of the same FPGA vendor. Thus, Fig. 4.17 is provided for example purposes only; the reader must consult the data sheets for FIFO blocks offered by a particular FPGA vendor when incorporating a FIFO into a digital system.

Some FIFO blocks have additional status signals named almost_empty and almost_full with configurable thresholds for these conditions. These signals are useful for assisting with controlling the data flow between the writing and reading digital systems. Two error conditions associated with FIFOs are:

- Writing to a full FIFO (input data is typically discarded). This condition is avoided by writing to the FIFO only when the `full` signal is negated.

- Reading from an empty FIFO (output data is unknown). This condition is avoided by reading from the FIFO only when the `empty` signal is negated.

In some FIFO implementations, the triggering of these error conditions may corrupt the internal FIFO status and produce erratic subsequent behavior, and error status signals (`read_error`, `write_error`) may be provided for system monitoring.

4.5   DUAL-PORT MEMORY

A dual-port memory has two ports, A and B, which support independent memory operations on each port. Figure 4.18 shows a typical interface for a dual-port memory. A dual-port memory that allows independent clocks for each port is sometimes referred to as a *true* dual-port memory.

Simultaneous operations to different memory locations have no timing constraints in relationship with each other. However, simultaneous operations to the same memory location will have timing constraints that vary by FPGA vendor. A typical specification for simultaneous access to the same memory location for a true dual-port memory is as follows:

- Simultaneous read access to the same location has no timing constraints.

- Simultaneous write operations to the same location produces unreliable data in that location.

- Simultaneous write and read operation to the same location produces correct data written to the location, but the read operation returns unreliable data.

The digital system designer using a dual-port memory is responsible for creating a system that avoids forbidden simultaneous operations. This usually involves external handshaking signals that coordinate access to the memory (the FIFO's empty/full signals fulfills this purpose in a FIFO design). Figure 4.19 shows two datapaths using a true dual-port memory and two handshaking signals, request (`req`) and acknowledge (`ack`), to send data from datapath A to datapath B. Figure 4.19a uses a two-phase protocol for accomplishing the data transfer; a change in the `req` signal indicates data availability from datapath A, with a corresponding change in the `ack` signal acknowledging receipt of the data by datapath B. In a two-phase protocol, data is transferred on each low-to-high transition of

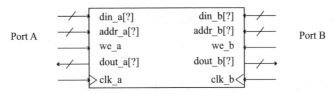

FIGURE 4.18: Dual-port memory.

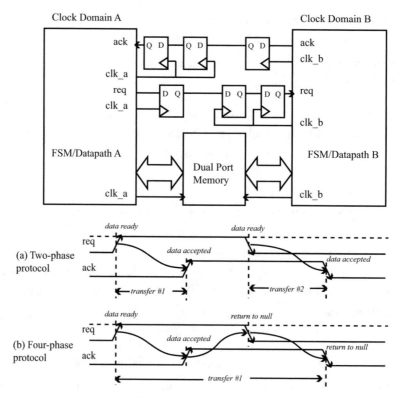

FIGURE 4.19: Dual-port memory use with handshaking.

the req line. A two-phase protocol requires changes in the req line to be detected, and is sometimes referred to as an edge-triggered or transition-sensitive protocol.

A four-phase protocol is used in Fig. 4.19b for accomplishing the data transfer; a logic one for req indicates data availability while a logic one for ack indicates data acceptance. Both the ack and req signals are negated (logic zero) before beginning a new data transfer. A four-phase protocol is referred to as a *level sensitive* protocol because the logic state of the handshaking signals indicate data availability and data acceptance.

Both four-phase and two-phase protocols can be readily expressed in modern HDLs. Some of the conventional pros/cons of two-phase versus four-phase protocols are as follows:

- A two-phase protocol requires more complex logic.
- A four-phase protocol maximizes signal transitions and thus energy consumed by those transitions.
- The return-to-null waiting period for the four-phase protocol may slow data transfers if the communication channel delay is long.

However, all of these pros/cons are technology and design dependent, with designer experience determining the protocol choice for a particular design.

The reader may question the necessity for using `req/ack` signals and instead want to indicate data availability by having datapath A write a nonzero value to a specified memory location being monitored by datapath B. This works only if the dual-port memory supports a simultaneous read during write operation to the same location, which is not the case for most true dual-port memories. It should be noted that if the two datapaths and the dual-port all share the same clock, then a simultaneous read during write operation to the same location is typically supported.

The advantages of a dual-port memory over a FIFO are that the dual-port allows bi-directional transfers between two datapaths and provides greater flexibility in data access. The disadvantage is that handshaking signals for avoiding forbidden simultaneous accesses may need to be provided by the designer.

4.6   ASIDE: SYNCHRONIZATION

In Fig. 4.19, the two DFFs clocked by `clk_a` on the `ack` input to datapath A and the two DFFs clocked by `clk_b` on the `req` input to datapath B are known as *two-flop synchronizers*. This is an accepted method for reducing the risk of an asynchronous input to a datapath input entering a *metastable* condition, in which the signal's voltage is stuck between a logic zero and logic one for an indeterminate period of time. A metastable condition can be triggered by a DFF's input failing to meet t_{su} and t_{hd} of the flip-flop. The probability of entering a metastable condition depends on many factors, some of which are:

- the internal design of the flip-flip
- the frequency at which the input signal changes
- the clock speed of the receiving system

A synchronizer is needed for any asynchronous input to a synchronous system. The reader is referred to [1] for a more complete discussion of metastability and synchronizer design.

In Fig. 4.19, the DFF clocked by `clk_b` on the `ack` output of datapath B and the DFF clocked by `clk_a` on the `req` output of datapath A are included to ensure that the `ack` and `req` outputs are glitch-free, that is, they only experience a single high-to-low or low-to-high transition during any clock period. These DFFs can be removed if these signals are already registered within the datapath. An FSM output signal that is generated by combinational gating using an FSM's state registers may experience glitches due to different delay paths through the logic gates. Because the `req` and `ack` outputs are asynchronous inputs to the receiving datapaths, these glitches could be treated as valid inputs, causing incorrect operation. If the two datapaths shared a common clock, then glitch-free outputs would not be needed because it is assumed that the outputs would be stable (satisfy t_{su} /t_{hd}) by the time the active clock edge occurred.

4.7   SUMMARY

This chapter has introduced the reader to commonly available embedded memory blocks found in modern FPGAs. Synchronous RAM blocks are preferred over asynchronous RAMs blocks because timing constraints for the designer are simplified when using synchronous RAM. Typical usage of RAM blocks requires counters to drive address lines, adding an extra clock cycle of latency from assertion of counter input to RAM output. FIFOs and dual-ports are useful for data exchange between datapaths that use different clock domains.

4.8   SAMPLE EXERCISES

1. Implement the datapath of Fig. 4.10 and ASM of Fig. 4.11 in the FPGA/HDL of your choice.

2. Modify the ASM of Fig. 4.11 to operate correctly if the registered dout output of the synchronous RAM of Fig. 4.10 is used instead of the unregistered dout output.

3. Compare the unregistered clock-to-dout time to the registered clock-to-dout time for an embedded memory block in an FPGA of your choice.

4. Using an FPGA of your choice, explore the timing characteristics for a FIFO that supports independent read and write clocks. Set the read clock to have 2/3 of the period of the write clock.
   a. How many read clock cycles does it take for the empty flag (read port side) to be negated when a write is performed?
   b. How many write clock cycles does it take for the empty flag to be asserted (write port side) when a read is performed that empties the FIFO?
      Repeat 4a, 4b with the read clock having 1/3 longer clock period than the write clock.

5. Using an FPGA of your choice, use an N-element FIFO with independent read/write clocks to create a design with the following characteristics:
   a. Set the FIFO size to be N-elements (your choice). Set the write clock to be 1/3 the period of the read clock.
   b. Create a write-side FSM that writes $2*N$ elements (use dummy data) to the FIFO at one write clock cycle per datum when a start input is asserted. Monitor the full signal to ensure that a write is not done to a full FIFO. Suspend writing if full is asserted; resume writing when full is negated. Halt operation when $2*N$ elements have been written to the FIFO.

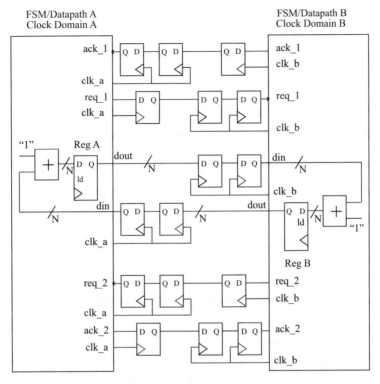

FIGURE 4.20: Asynchronous transfer.

    c. Create a read-side FSM that removes elements from the FIFO whenever the empty signal is negated; remove data as fast as possible from the FIFO (one clock per datum). Ensure that your FSM does not attempt to read from an empty FIFO.

    d. Change the read/write clocks such that the write clock has a 1/3 longer period than the read clock. Verify that your design performs as expected.

6. This problem refers to Fig. 4.20. Using four-phase handshaking and with datapath A clock 2/3 the period of datapath B, create FSMs for dapathpaths A/B that accomplish the following (steps *a* through *c* are FSM A operation, steps *d* through *f* are FSM B operation).

    a. After reset, FSM A initializes Register A to zero.

    b. FSM A then transmits the Reg A value to FSM B using the handshaking pair req_1/ack_1 and its dout bus.

    c. FSM A then waits for a value to be transmitted back from FSM B on its din bus and using the handshaking pair req_2/ack_2. This new value is incremented by 'one' via the adder, and loaded into Reg A(at this point, FSM A loops through steps *b* and *c*, resulting in a continuously incrementing value being transmitted between FSM A and FSM B.)

    d. After reset, FSM B initializes Register B to zero.

    e. FSM B then waits for a value on its din bus to be transmitted from FSM A using the handshaking pair `req_1/ack_1`. This value is then incremented by 'one' via the adder, and loaded into Reg B.

    f. FSM B then transmits the Reg B value to FSM A using the handshaking pair `req_2/ack_2` and its `dout` bus (at this point, FSM B loops through steps *e* and *f*, resulting in a continuously incrementing value being transmitted between FSM A and FSM B.).

7. Repeat problem #6 using two-phase handshaking.

8. Using the FPGA of your choice, create a dual-port memory design similar to Fig. 4.19 that has the following characteristics:

    a. Set the datapath A clock to be 1/3 the period of the datapath B clock. Use a four-phase handshake protocol to coordinate access to the dual-port.

    b. Using the initialization mode of Fig. 4.8 as a guide, have datapath A write the value N to location zero of the dual-port and then the data to be summed into locations one through $N + 1$. Once the dual-port has been initialized, have datapath A inform datapath B that data is ready to be summed through the handshaking protocol.

    c. Have datapath B read location zero to determine the N value, then sum the values in locations 1 through $N + 1$. Once datapath B is finished, use the handshaking protocol to inform datapath A that the data in the dual-port has been consumed, and then resume waiting for another data packet to be placed in the dual-port by datapath A.

9. Repeat problem #7 using the two-phase handshaking protocol.

4.9 PROJECT SUGGESTION

The latter part of Chapter 3 used a FIR digital filter to explore issues in datapath scheduling. The general form of an N-order FIR digital filter is:

$$y = x \times a0 + x@1 \times a1 + x@2 \times a2 + \ldots + x@N \times aN \qquad (4.1)$$

The x value represents the current input sample value, $x@1$ the input sample value from the previous sample period, $x@2$ the input sample value from two sample periods previously, etc. The filter coefficients $a0$, $a1$, ... aN determine the filter's performance characteristics such as low pass, high pass, band pass, etc. A JAVA applet that produces FIR filter coefficients given a filter specification is available at [2]. Typical results from the applet are given in Table 4.1.

    This project's task is to build a fixed-point, *programmable* FIR filter that allows the filter order and coefficients to be dynamically loaded. As with the memory sum example of Section 4.3, the filter has an initialization mode in which the filter order and coefficients are loaded, and a

TABLE 4.1: FIR Filter Example

Rectangular window FIR filter, Filter type: Low Pass (LP), Order: 20

Passband: 0 – 1000 Hz, Transition band: 368 Hz, Stopband attenuation: 21 dB

Coefficients:

a[0] = 0.00360104 (0x007)	a[11] = 0.230304 (0 x 01D7)
a[1] = 0.027779866 (0x038)	a[12] = 0.13769989 (0 x 011A)
a[2] = 0.032870565 (0x043)	a[13] = 0.03300727 (0 x 043)
a[3] = 0.009205259 (0x012)	a[14] = -0.03924712 (0 x FAF)
a[4] = −0.030985044 (0x0FC0)	a[15] = −0.057350047 (0 x F8A)
a[5] = −0.057350047 (0xF8A)	a[16] = −0.030985044 (0 x 0FC0)
a[6] = −0.03924712 (0xFAF)	a[17] = 0.009205259 (0 x 012)
a[7] = 0.03300727 (0x043)	a[18] = 0.032870565 (0 x 043)
a[8] = 0.13769989 (0x011A)	a[19] = 0.027779866 (0 x 038)
a[9] = 0.230304 (0x01D7)	a[20] = 0.00360104 (0 x 007)
a[10] = 0.26717955 (0x223)	

computation mode that accepts new input samples and produces a new output value for each input sample. Figure 4.21 gives the cycle specification for initialization mode, which is entered when start is asserted and mode is a logic one. The start input is negated when the last filter coefficient is entered.

In Fig. 4.22, computation mode is entered when start is asserted and mode is logic zero. The filter then waits for assertion of input ready (irdy), which indicates that a new sample value is present on the din input data bus. The filter asserts output ready (ordy) when the filter computation is finished and the dout data bus contains the final result. The filter then returns to waiting for the next assertion of irdy. Computation mode is exited when start is negated.

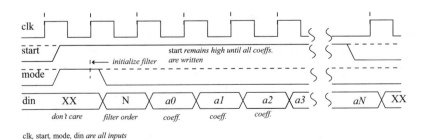

clk, start, mode, din *are all inputs*

FIGURE 4.21: FIR filter initialization cycle specification.

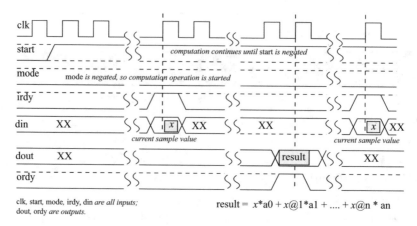

clk, start, mode, irdy, din *are all inputs;*
dout, ordy *are outputs.*

result = $x*a0 + x@1*a1 + ... + x@n * an$

FIGURE 4.22: FIR filter computation cycle specification.

4.10   IMPLEMENTATION HINTS: SIGNED FIXED-POINT, EXAMPLE DATAPATH

The coefficients of Table 4.1 include negative values, so one choice for number representation is two's complement fixed-point representation (unsigned fixed-point number representation was explored in Chapter 3). Given N bits, two's complement represents the integer range $2^{N-1} - 1$ to -2^{N-1}. For example, 12-bit 2's complement represents the integer range $+2047$ to -2048. This range can be mapped to the number range $(+1.0$ to $-1.0]$ by dividing each integer by $+2^{N-1}$. A fractional value in the range $(+1.0$ to $-1.0]$ can be mapped to its binary value by multiplying it by 2^{N-1}. The range $(+1.0$ to $-1.0]$ is a good choice for a fixed-point digital filter implementation because the output of an unsigned N-bit analog-to-digital converter (ADC) that samples an analog input is easily converted to this range by subtracting 2^{N-1} from the ADC output code. The hex values given for the coefficients of Table 4.1 are the 12-bit two's complement representations calculated by multiplying each coefficient by 2048.

Fig. 4.23 shows an example datapath for implementing the programmable filter. Input samples are assumed to be two's complement 12-bit, mapped to the range $(+1.0$ to $-1.0]$. Two single-port RAMs are used for storing the coefficients and previous input samples.

The movement of the counters that address the sample and coefficient RAM during the calculation for a single input sample $x0$ is shown in Fig. 4.24. The coefficients are stored in the first $N + 1$ locations of the coefficient RAM, in order from $a0$ to aN. The $N + 1$ sample values used in a calculation ($x0$ through xN) are stored in the first $N + 1$ locations of the coefficient RAM, but the samples values are stored in decreasing memory locations from wherever the current sample $x0$ is stored (this is because arriving samples are stored in increasing memory addresses, so decreasing memory addresses contain *past* input samples).

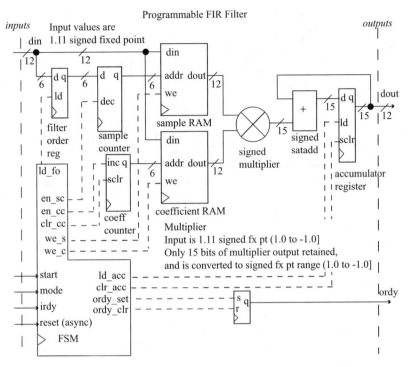

FIGURE 4.23: Sample datapath for FIR programmable filter.

Because the datapath contains only one multiplier and one adder, an FIR calculation for a new input sample requires at least $N + 1$ clocks. The multiplier is a *signed* multiplier, which is generally available as a building block from FPGA vendors. It was mentioned in Chapter 3 that a K-bit $\times K$-bit multiplier produces a $2K$-bit result. For unsigned fixed-point numbers mapped to the range $(1.0 - 0.0]$, it was noted that the lower K-bits of the $2K$-bit product could be discarded, since these represented the K least significant bits, and the datapath size could be kept at K-bits.

However, what bits should be discarded for a signed K-bit $\times K$-bit multiplier using numbers in the range $(+1.0$ to $-1.0]$? One may intuit that it would also be the least significant K-bits, but the true answer is somewhat more complex. To illustrate, examine Eq. 4.2 that shows the multiplication of $+0.5 * -0.5$:

$$y = (+0.5) \times (-0.5) = -0.25 \tag{4.2}$$

The numbers $+ 0.5, -0.5$ mapped to 12-bit two's complement are $+ 0.5 * 2048 = 1024 = 0\times400$ and $- 0.5 * 2048 = $ -1024 $= 0\times C00$. The signed binary multiplication of Eq. 4.2 produces:

$$y = (0x400) \times (0xC00) = 0xF00000 \ (24 - \text{bit product}) \tag{4.3}$$

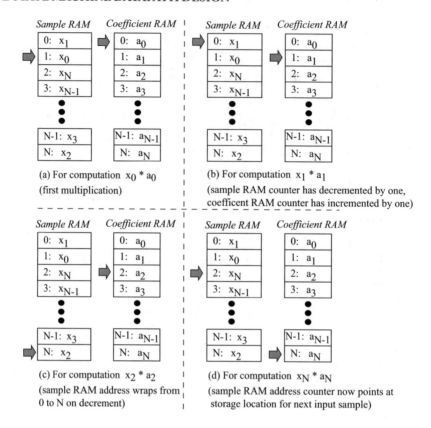

(a) For computation $x_0 * a_0$
(first multiplication)

(b) For computation $x_1 * a_1$
(sample RAM counter has decremented by one,
coefficient RAM counter has incremented by one)

(c) For computation $x_2 * a_2$
(sample RAM address wraps from
0 to N on decrement)

(d) For computation $x_N * a_N$
(sample RAM address counter now points at
storage location for next input sample)

FIGURE 4.24: FIR computation.

Dropping the least significant 12-bits (last three hex-digits), the value 0 x F00 is equal to -256 as a 12-bit two's complement integer. Mapping -256 to the range $(+ 1.0$ to $- 1.0]$ produces:

$$-256/2048 = - 0.125 \qquad (4.4)$$

which is one-half the expected value of $- 0.25$. Equation 4.5 shows the reason for this by examining the number range of the multiplication result:

$$(+1.0, -1.0] \times (+1.0, -1.0] = (+2.0, -2.0] \qquad (4.5)$$

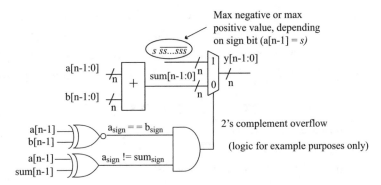

FIGURE 4.25: Two's complement saturating adder.

The multiplier output range has to be extended by an additional integer bit because the value $+1.0$ is now included in the output range (because $-1.0 * -1.0 = +1.0$). This means that the upper two bits of the 24-bit product are dedicated to the sign and integer portion of the result. This also has the unfortunate result that the output number range of $(+2.0, -2.0]$ is now different from the input number range of $(+1.0, -1.0]$. The extra bit needed for the integer portion of the product to encode $+1.0$ is wasted if the multiplier is never given the inputs of $-1.0 * -1.0$. Because one of the multiplier inputs is always a coefficient, the coefficient choices can be restricted to not include -1.0. This means that actual range of values produced by the multiplier fall in the range $(+1.0, -1.0]$ and thus the most significant bit of the multiplier can be discarded. Note that discarding the most significant bit is the same as shifting the multiplier output to the left by one, which is multiplication by two. Multiplying the result of eq. 4.4 by two gives the expected result: $-0.125 * 2 = -0.25$.

The datapath of Fig. 4.24 shows 15 bits of the 24-bit multiplier product being retained (nine bits are discarded). The bits discarded from the 24-bit product are the most significant bit, and the eight least significant bits. This gives three extra least significant bits for rounding purposes as the FIR sum is being accumulated. Only the most significant 12-bits of the accumulator register are used for the dout output result.

The adder shown in the datapath of Fig. 4.24 is a two's complement saturating adder, which saturates the output result to the maximum positive or maximum negative value if two's complement overflow occurs. Fig. 4.25 shows a conceptual implementation for a two's complement saturating adder (this logic works but more optimal implementations exist).

4.11   TESTING THE PROGRAMMABLE FILTER

One easy method of testing the filter is to apply an input sample of -1.0, followed by zeros. This produces output values of $-a0, -a1, -a2, -a3, \ldots -aN, 0, 0, 0$, etc. By implementing the FIR filter function in a programming language of choice, any arbitrary numerical input stream can be provided and the resulting output stream of the implementation is checked against expected results.

An optimum check is to provide a digitized sine wave of a particular frequency and observe the output to determine if the filter function (low-pass, high-pass, band-pass) is accomplished. The psuedo code in Listing 1 produces input values for one cycle of a sine wave for a given frequency f sampled at a frequency of S (the digital filter applet of [2] assumes a sample frequency of 8000 Hz).

Listing 4.1: PSUEDO-CODE FOR DIGITIZED SINE WAVE

```
// f is sine wave frequency (Hz)
//S is sampling frequency of the filter (Hz)
for (t = 0, j = 0; j < (2 * π); t++, j = (t*f*2*π)/S) {
x = sin(j); //x is input sample value
}
```

Fig. 4.26 shows a sine wave input to a 20 tap LP FIR filter with a cutoff frequency of 100 Hz. The input sine wave has several cycles at 100 Hz (the edge of the pass band), followed by several cycles at 300 Hz (in the filter's transition band), followed by several cycles at 600 Hz (in the filter's stop band). The output waveform shows attenuation as the input waveform's frequency increases, which is expected for a low-pass filter.

4.12   FILTER IMPROVEMENTS

Many alternatives are possible for the example datapath shown in Fig. 4.23.

- The coefficients of N-order FIR filter are symmetric as seen in Table 4.1; $a0 = aN$, $a1 = a(N - 1)$, etc. The number of memory locations used in the coefficient RAM can be reduced from $N + 1$ to $(N/2) + 1$.

- The number of clock cycles required for producing the output given an input sample can be reduced by distributing the input samples and coefficients among multiple RAMs and including more multipliers and adders. This is the hardware resource versus computation time tradeoff examined in Chapter 3.

- The maximum clock period can be decreased at the cost of greater clock cycle latency by using the registered dout output of the RAM blocks and by placing a pipeline register between the multiplier and adder.

- Some FPGA vendors offer embedded RAM blocks that have built-in shift register function-ality as required for digital filter implementations and could replace the counter logic that is currently used to access the RAMs.

- Some FPGA vendors offer library support for floating-point execution units; change the datapath from 12-bit fixed-point to single-precision floating-point.

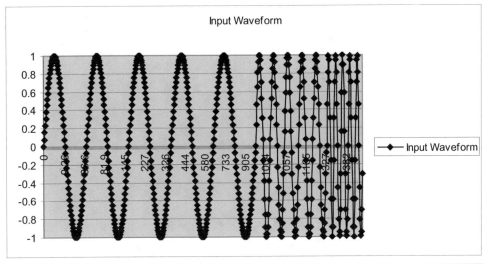

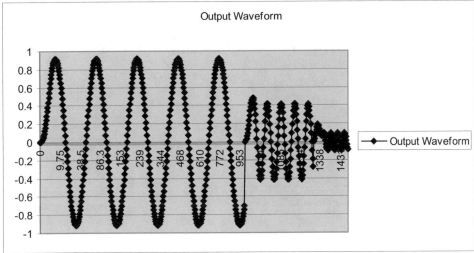

FIGURE 4.26: Filter input versus filter output.

4.13   REFERENCES

[1] R. Ginosar, "Fourteen ways to fool your synchronizer", *Proc. of the Ninth International Symposium on Asynchronous Circuits and Systems*, 12-15 May 2003, pp 89-96.

[2] FIR Digital Filter Design Applet, Online as of August 2007: http://www.dsptutor.freeuk.com/FIRFilterDesign/FIRFilterDesign.html.

Author Biography

Justin Stanford Davis received his Ph.D. in Electrical Engineering from the Georgia Institute of Technology in August 2003, as well as his M.S. and B.E.E. degrees in 1999 and 1997. During the summers of 1998 and 1999, he worked at Hewlett-Packard (now Agilent Technologies). In fall of 2003 he joined the faculty in the Department of Electrical Engineering at Mississippi State University as an Assistant Professor. In the summer of 2007 he joined Raytheon Missile Systems as a Senior Electrical Engineer. His research interests include digital design for high-speed systems, SoCs, and SoPs, as well as signal integrity and systems engineering.

Robert B. Reese received the B.S. degree from Louisiana Tech University, Ruston, in 1979 and the M.S. and Ph.D. degrees from Texas A&M University, College Station, in 1982 and 1985, respectively, all in electrical engineering. He served as a Member of the Technical Staff of the Microelectronics and Computer Technology Corporation (MCC), Austin, TX, from 1985 to 1988. Since 1988, he has been with the Department of Electrical and Computer Engineering at Mississippi State University, Mississippi State, where he is an Associate Professor. Courses that he teaches include VLSI systems and Digital System design. His research interests include self-timed digital systems and computer architecture.

Printed in the United States
by Baker & Taylor Publisher Services